Pulsars- An Adventure to Explore the Wonders of the Stars…

P.A. Murad

Pulsars- An Adventure to Explore the Wonders of the Stars…

P.A. Murad

ISBN-13: 978-1515273851

ISBN-10: 1515273857

© Copyright 2015, Paul A. Murad, Artslogo LLC. Printed by permission.
All rights reserved. No part of this book may be reproduced, stored in a retrieval system, or transmitted by any means, electronic, mechanical, photocopying, recording, or otherwise, without written permission of the author. The names are fictitious and no reference to actual people, alive or dead, is intended.

"The views expressed in this manuscript are those of the author and do not reflect the official policy or position of the Department of Defense or the U.S. Government."
P. A. Murad

Preface

It is amazing at how much mankind has learned, grown and developed regarding the ability to understand some things about the cosmic environment. We seem to feel that with embryonic technological innovations and satellites, we are discovering new answers to new questions. But in reality, we only discover more questions when we suddenly find more answers in a never-ending cycle. This must be growth. Mankind needs to acquire an intellectual effort in an attempt to satisfy his curiosity and to understand developing faster than light propulsion schemes. If we cannot achieve faster than light travel, then mankind is doomed to marooning its existence on the big blue marble we affectionately call the Earth. This is obviously an ambitious undertaking.

Part of this effort is to explore some small portion of a very difficult problem by examining different disciplines. These efforts would provide the necessary ingredients and reveal new insights. The difficult problem is to travel at faster than the speed of light. Pulsars possibly fall within this category because of some unusual behavior to understand gravity. For example, several individuals did excellent work on characterizing a specific binary pulsar with very unusual orbital characteristics. The bizarre orbital behavior suggested how to account for the energy loss. The solution was that the neutron star radiated considerable amounts of gravitational waves. Although they performed considerable intellectual efforts winning the Nobel Prize, no information about the details regarding the gravity wave details such as its frequency, spectrum, or waveform is not described anywhere. We have to ask the obvious question if it could have been some other effect in lieu of gravity waves that may perhaps account for the energy loss elsewhere.

Pulsars were unknown until about the late sixties using radio telescopes. Basically these pulsars created a sweep of an electromagnetic radiation swath located at poles. These sweeps would cover over part of the surface of the Earth. This pulse-like signal would occur for a period enduring several times per second. Moreover, this process acted similar to a lighthouse sweeping a region with a response that was accurately repeatable, periodic and obviously represented an unusual cosmic endeavor.

With the developments of the Hubble Telescope as well as other satellite systems, we have a better idea about these unusual phenomena. Astronomers looked at this behavior and tried to characterize these events to understand its importance. These first images started to increase the intellectual investment process to provide some sort of pictures or digital images of what was going on with Binary pulsars.

Imagine if we can describe these very far away events where we could not see with our own eyes! Still, it is amazing these things can dance and perform a very unusual reverie. It attracted me based upon the unusual orbits they produced. In these neutron stars with their companions in a binary, they seemed to be married in unusual orbits. These characteristics suggest there is a unique relationship that depends upon the masses of the two bodies and the very high spin rate of the neutron star itself. These ingredients generate unusual 'free' orbits. For the success of this marriage, the neutron star had, in my perspective, would generate a predetermined rotation rate to imply some connectivity between gravitational attractions with a side effect due to angular rotation. These ingredients were the causes and the subsequent orbits were the effects that tantalized my motivation.

To my way of thinking, this seemed to reveal what one might suggest is that the pulsars response produced a unique set of orbits. I had studied and was aware of unusual gravitational laws about losing weight with the neutron star's rotation speed that might have been ordained for a neutron star within a binary. Moreover, other gravitational laws implied gravity was not only an attractive force but also included angular momentum. Both of these characteristics seemed to belong to the neutron star's performance in this dance within the binary. This was very unusual and these activities are usually outside of the scope and interest of the typical engineer or scientist.

In other words, binary pulsars may reveal experimental evidence of some unusual gravitational abilities that could put forth information necessary to travel in the far-regions of the cosmos. This foreknowledge about any difference in gravity might be the tenets to develop a far-reaching space propulsion scheme to reach the speed of light.

Now basically there is a differences regarding engineers and physicists. Engineers use a constant value for gravity and have been considerably successful at this to predict satellite and planet orbits as well as send out long-range vehicles to the Moon or even to venture out to the far cosmos of the solar system. In contrast, the physicist uses Einstein's field equations and compels defining a gravitational tensor. This starts with the assumption where the main diagonal is real and off-diagonal terms generally vanish. Interestingly, the physicists respond to the same value as the engineer but only after considerable mathematics.

The problem is that physicists enjoy to think in a symmetric fashion. We all do. What is not considered is if the off-diagonal terms did not vanish, it could produce significant gravitational effects that may be achieved with what we are suggesting with unusual events. This could include gravity with attraction and producing angular momentum. If this occurred with further research, this would definitely have space propulsion implications. However, the physicists claim this problem would be too difficult to solve using off-diagonal terms in the field equations!

Further, physicists' claim Einstein's field equation requires a hyperbolic or wave partial differential equation but still continue to use Newtonian gravitation. The latter does not include time and does not allow the existence of gravitational waves.

There was also another problem. In any basic dynamics course, the problem of a satellite moving around the Earth is given. This simplified problem is the *'captured'* orbit situation where large differences exist between the masses of the two bodies usually the Earth and a small satellite. We are not taught to teach the problem where you have two objects of almost near masses and these are in *'free'* orbits where these masses may not coincide on the same orbit. At this point you wonder if there was a hidden body at the focal point common to both orbits somewhat similar to the Sun within our solar system. Imagine how the trajectories would occur for all of the planets if the Sun suddenly disappeared? What was going on? Was there a gravitational vortex or a gravitational singularity at the focal point?

Dr. Robert M. L. Baker, Jr. a colleague mentioned the other orbits with similar masses would also result in elliptical orbits and this was previously established. I could not find such references to what I call the 'free' orbital problem. On this basis, I had to understand this possibility to also examine some of the previous comments about unusual gravitational laws. This is also considered in this monogram and this analysis was quite enjoyable in so doing.

I would like to offer an interesting point for the audience. I once studied Baker's textbook in 1967 as a graduate student at NYU to study astrodynamics. During 2003, he personally asked me to put together a High Frequency Gravitation Wave Conference, which I was fortunate to do. Several years later, we co-authored several papers vis-a-vis parallel universes and other topics regarding gravitation. I have to mention this was all enjoyable and unexpected during my career; it only demonstrates that it is very difficult to predict the future generally about anything. Imagine reading a technical book and then meeting the author as a co-author more than 40 years later!

With these interests, I decided to write this monogram. Each chapter consists of technical papers presented in peer-reviewed papers or conferences. The first chapter covers some notions about Einstein's field equations. Unfortunately this is not used to actually solve these equations related to Binary Pulsars but rather to expose the engineer to these notions. Once understood, this may provide an opportunity to use this powerful tool to solve engineering problems and possibly develop a warp drive in an engineering perspective. The second chapter covers the problems that address the motivating factors, which caused my initial interest. These trajectories were unusual in the basic knowledge about orbits. The third chapter covers notions about how neutron stars might be formed and how these notions are extended to possibly develop a future space Propulsor. The fourth chapter looks at the 'free' orbits of these binaries, a neutron star and companion, which are confined in these specific orbits. This solution is considered somewhat commonplace, but even here are some interesting insights about the binary trajectories designed to resolve some of the concerns covering the preceding chapter.

P.A. Murad
Morningstar Applied Physics, LLC. Virginia, USA

[a] CEO, pm@morningstarap.com.
Copyright © 2015 by the author. Printed by permission.

Foreword

There are many wondrous events that require understanding various cosmic issues. We should always keep an open mind and examine all possible venues to discover anything new or interesting.

One of these is the hypothesis of the Big Bang. On certain scientific issues, individuals choose to raise issues about the presence of God or not being the cause on or about the Big Bang. I would prefer to simply address some points to clarify my humble understanding about the Big Bang. We try to understand some of our basic and complicated notions which could be an event that is far greater or yet simpler than our own perception. Likewise, our perception with assumptions may be flawed. For example, when the Big Bang starts, mass is created. By conservation, somehow we have to offset the created mass. To balance this mass, this is assumed to be equal with dark matter. I find this intriguing because we think of everything as if there is a huge scale in space somewhere which must be balanced because if it does not, then there is a clear scientific violation.

Dark matter was originally discovered to counteract for an explanation for the mass in the cosmos as well as the differences in explaining the accelerations due to the formation of galaxy's spirals and interactions with other bodies. These responses tend to imply that Newtonian gravitation does not work for these conditions. The thought is if this dark matter exists, this new mass offsets gravitation to account for the differences. And since we cannot see it, the matter must be dark. One wonders if the gravitational model based upon Newtonian gravitation, may be inadequate especially if we look at regions far away from our own solar system. Here, we think we understand this phenomenon of gravity within our own solar system. However, things might change because we just do not know or have any experimental evidence observing the influence of gravity outside of our near-field and bounded solar system. Moreover, some claim the charge of dark matter has no charge. If we look at conservation, somehow someone ignored the conservation of charge because they violated the use of this scale.

One would wonder if you are in a spaceship moving at the speed of light or greater, can it use some fields to move obstacles away to deflect normal matter. What happens if we contact with dark matter to move it out of the way from the path of a moving spacecraft? If dark matter has no electrical or magnetic charge per the assumption, this possibility does not exist to move the dark matter aside, then it implies there would be structural damage in a collision with our fast moving spaceship along the way.

Could it be that this artificiality is false? The other views are that maybe another gravitational law should be established to examine these anomalies and include these premises within the conventional wisdom? The other point regarding reducing gravity, centrifugal motion tends to counteract gravity and this rotation could provide the rationale for these differences from our expectation of Newtonian gravitation.

The Big Bang theory suggests we also crave to define the creation of energy to account for what enters into our four-dimensional space-time continuum. When this occurs, before the Big Bang, there is presumably no matter, no energy and no time. One obvious question is if there was space with the beginning before the Big Bank to the present after 13.7 billion years, how could you make a statement occurring prior to this event, there was no time? It must have been a void of emptiness forever. One could ask how long the time period of empty vacuum was, but then I digress. This is once again a concern about the conservation of the energy from all of dark matter. Hence we make an artifact where this is also created by dark energy.

As an aside, one wonders about conservation of momentum. The point is maybe the big bang was not an explosion but acted like a 'big' faucet from another dimension or dimensions pouring forward all of this mass and energy into our space-time continuum. Dare I say this is caused by God? If it did, then the faucet would need to conserve momentum where there would be some large concentration of matter in one particular direction like the water issuing forth from a faucet. This difference in momentum should be observable but what would the signatures be? With the Big Bang, there is no alarm about momentum because the explosion is assumed unidirectional from a central source. Or is it? This would mean the entire cosmos must have initially been spherical like an explosion if you could only find the edges of the cosmos. Is this possible without our technical capabilities? If there was a 'great' faucet, the edges of the cosmos should have an unsymmetrical appearance and we, as usual would have demanded the existence of dark momentum.

We have discussed dark matter and dark energy. One thought appears is that these characteristics would be different from, say matter or energy as we currently understand. For example, when the Big Bang started, maybe it was coated with only dark matter/energy where the center of this ball has normal matter and normal energy in the center. Thus, the dark matter and dark energy would be propelled to the furthest of the cosmos from the rest of normal matter and energy. On this basis, there would be no interaction and we go on only concerned how the initial assumption about a coated sphere before the Big Bang explosion is realistic or not. The thought is regardless of all of these characteristics, they all probably have to interact with electricity, magnetism, gravity and let us call a fourth force for completion called torsion. On this basis, if normal matter can be charged, it should be likewise familiar where dark matter should also be charged but in a different way. All of these depend upon these four forces. Since we do not as of yet understand how this torsion field would interact with normal matter or energy, we really cannot make any judgment about how this torsion would similarly react with dark matter or dark energy.

If there is either a Big Faucet or a Big Bang, how can we not accept the premise that the cosmos is expanding? The faucet would inject in one specific direction with initial velocity and the Big Bang as an explosive would move an initial velocity in a radial direction at the center of the explosion. This should be only a minor consideration. Why would one suppose a contraction occurs for some time period, change, and then continued at the current expansion rate? This to me, is due to momentum despite the gravitational attraction between these expanding celestial bodies. At any point in these questions, someone could ignore the presence of God or that God does not exist. I feel I have to make a personal expression about the response to this point. There is a feeling about this answer but I do not wish to force my views on someone else. Let us call this as just another one of the mysteries of our cosmos if we wish to explore and look forward to understand with no clear explanation about the events around our cosmic neighborhood.

This theology of the Big Bang has one more poor assumption concerning our views regarding conservation in both mass and energy. The problem is if you believe in the Theory of Relativity, Einstein's field equation is unusual and implies that if mass, momentum, and energy conservation are violated, they are used to create space-time curvature or change gravity within our space-time continuum. If there is a torsion field, we also have to assume this violation will also change the torsion field. In other words, our views about dark mass and dark energy need to be reconsidered. We also crave to break this childish paradigm to feel conservation, although extremely valuable for normal circumstances, may have some limitations where we need to explore if we wish to find and understand new physics. Maybe this new physics relies purely upon none-conservation.

Let us spend some time about theology. Obviously if there was a giant faucet, one could argue God's existence is there to turn on the faucet. You could also make comments about God's presence with the Big Bang, but the issue is we are clearly mired with difficulties with Mother Nature as well as we might think about how God would react. For example, many different cultures talk about God's garden with the first man and the first woman.

Let us consider Adam and Eve. They are the gardeners but they cannot see except after they eat the apple from the tree of life. The tree of life where they cannot touch is right in the middle of the Garden. If it was so evil, why didn't God place it of on the corner of the Garden or even outside of the Garden, somewhere away from the center of focus? In reading Genesis, we assume these events are really fast paced. Maybe these two people were very obedient and they obeyed God for, let's say, maybe years, decades, or even centuries. Sooner or later, the human curiosity would set off these two individuals and they will do something wrong to explore the nature of the tree of life. This leads them to get exiled from the Garden and separated from God.

The point is that the deck may have been stacked against these two naïve individuals. What kind of gardeners could they be if they were blind or did not see? How could they tell the difference between 'good' plants and fruits from those of weeds? Was God pleased with their work? Finally, the tree would create such tension and inquisitiveness to mankind where we had to eventually, even after centuries, disobey God because of our innate curiosity. The issue is God or Mother Nature just does not give us a break!

Let us assume mankind is bright. A Kaballalist asked a basic question after Moses some 500 years after he passed away. The question was: "What was the comments made between Moses and God?" Would this question be relevant? Why did the Jews, who are usually bright people, wait for so long to ask this question?

And what about Newton's bucket where you are sitting on a hill and looking at the stars. After some time, the stars move in the heavens. Did the stars move or did the Earth move? It took 300 years for Einstein to resolve this question where you need to use a fixed reference point with respect to the stars and the Earth. By looking at reactions of the reference point, you can resolve both move. The point in all of this is we do not ask the right questions and when we try to get an answer, it is not given to us very easily but grudgingly. There is much toil to uncover the mysteries of the cosmos. This is what we expect from scientific investigations.

Regarding cosmology, there is a panoply of items regarding alleged gravitational anomalies in the Solar system. Some of these are:

    a) Possible anomalous advances of *planetary perihelia*,
    b) Unexplained orbital residuals of a recently *discovered moon of Uranus*,
    c) The lingering unexplained secular increase of the *eccentricity of the orbit of the Moon*,
    d) The so-called *Faint Young Sun Paradox*,
    e) The secular decrease of the mass parameter of the Sun,
    f) Satellites that involve the *Flyby Anomaly*,
    g) The *Pioneer Anomaly*, and

h) The anomalous secular increase of the astronomical unit.

These issues obviously are crucial throughout each celestial body outside of our solar system because of our lack of understanding about gravity. Although interesting and there is some connection regarding pulsars, these subjects are briefly discussed; however, the focus of this effort is directed more toward neutron stars and pulsars. Evenso we still cannot discuss all there is to know about pulsars. For example, if a supernova can create either a neutron star or a black hole, what is the criteria for this difference to occur? Moreover, if one would assume that a companion in a binary pulsar, either with a neutron star that captures another neutron star or a celestial body, can a supernova create a binary pulsar with two neutron stars? Would the core of a neutron star be covered with another material such as iron to allow magnetic and electric fields to exist in lieu of using electron or proton gases? These are issues yet to be established.

To you that wish to pursue the adventures to discover the pulsars and neutron stars to hope and touch the face of God, your journey will not be easy and sometimes the journey itself may be more pleasure than reaching the final destination. Learn as much as you can and make contributions to share for mankind. Go out there and let us go slay a dragon!

P.A. Murad
Morningstar Applied Physics, LLC. Virginia, USA.

CEO, *pm@morningstarap.com.*
Copyright © 2015 by the author, permitted by permission.

Table of Contents

Preface   3

Table of Contents   7

CHAPTERS

1. Relativity, Reality and Relevance   15
2. Pulsar Behavior that may impact a Future Space Propulsor   41
3. An Assessment Concerning Neutron Stars and Space Propulsion Implications   57
   - Appendix A. Pulsar Timing Details   73
4. A Tutorial to Solve the 'Free' Two-Body Celestial Mechanics Problem   75
5. Relativistic Orbit for the 2-Body Celestial Mechanics Problem   91
6. Propulsion and Implications   105

    Epilogue   109

    Subject Index   113

Chapter I.

Relativity, Reality and Relevance

Abstract: The notion that one can go faster than the speed of light, travel in parallel universes, or even go back and forth in time is an interesting scientific conjecture. These activities require evaluation within Einstein's field equations from an engineering perspective to possibly create a suitable propulsion system. A set of circumstances may lead to an anomalous gravitational distortion in what is normally referred to as the "vacuum field equation" in a near flat space-time continuum with no electric, magnetic or torsion fields. This distortion, possibly created by a gravitational vortex, is driven purely by the gravitational tensor. Moreover, these field equations may imply that unusual consequences could follow if one includes nonlinearities, off-diagonal terms in the gravitational or space-continuum tensors, more detailed boundary conditions as well as creating systems that enjoy over-unity performance. Depending upon definition of the curvature tensor, conditions for travel in a parallel universe, faster than light travel and past/future time travel may exist. The problem is to incorporate these capabilities within a design of a suitable yet to be considered propulsor based upon anomalies that exist with pulsar behavior to explore unknowns within the cosmos.

Keywords: Gravitational vortex, space-continuum, magnetic fields, electric fields, torsion.

I. INTRODUCTION

If one thinks mankind is an intellectual paradigm, think about the thought throughout all of mankind, only Newton asked the question about gravity when the apple fell. There are many different views about Newton's bucket. If you sit on a late night on a hill and look at the stars, they rotate with some uniformity. The question is if an individual sits in a bucket, does the stars move, does the Earth move, or does he move inside the bucket?

No one answered this question for about another 300 years. The solution was resolved from Einstein where he suggested you can only answer this question if there is a known point of reference with location, and velocity (if you are brave, include acceleration). From this reference point, you can then decide that indeed the stars move slowly but the majority of the rotation is due to the Earth. Thus it is obvious we should have some understanding about Einstein's field equations.

I would like to also share some technical thoughts with you based upon a rather interesting daylong event which ended with the Premiere of the Sirius movie. There were several interesting comments worth repeating here.

Some views to make assessments would look at evaluating 'fake' designs or concepts by exploring and adhering to the conservation laws. If a new concept performs in an unusual fashion, you should look at the conservation laws to ensure this is not violated. Such notions indicate angular motion cannot produce linear motion. This may indeed be true but there are some problems especially when a device might exist that actually changes gravity. As you all know about the conservation equations, as engineers. We are very comfortable to do this but we have no terms allowing for changing gravity. Changing gravity does not impact continuity, not in momentum or in the energy conservation. Obviously, engineers cannot deal with changing gravity…

For several years, there exists an argument where engineers and physicists talk a different language despite the existence of Einstein's field equations which offers some level of unification in these different ideas and thoughts. Let us at least expand this and if views are wrong, feel free to identify this.

In spite of being relatively easy to suggest where a warp-drive propulsion system can be created, the system will be quite difficult to design and build. Before proceeding, we should examine Einstein's Field Equations of General Relativity [1]-[3] to determine if there are any peculiarities, which can give us various insights into such a venture. The notion one can go faster than the speed of light, travel in parallel universes, or even go back and forth in time is an interesting scientific conjecture. These activities require evaluation within Einstein's field equations from an engineering perspective to possibly create a suitable propulsion system. In addition, these issues should be examined with an open mind before restricting ourselves to notions such that it takes all of the mass within the solar system to accelerate a spacecraft to near light speed or to understand the dynamics of a neutron star. There simply has to be another path if mankind is to get off of this rock we call the Earth. Whether such an avenue is currently available depends obviously upon our technological maturity and this is really a subject not to be addressed here; we need to approach space travel on a more parochial level.

Let us look at this problem from a clean slate rather than cling to historical notions developed within the conventional wisdom. Yes, based upon the conventional wisdom, mankind would clearly be doomed to rot perpetually on the Earth and forever will be powerless to cavort amongst the stars. There are many problems of which the most serious one is humans cannot travel faster than the speed of light because the relativity factor becomes an imaginary number. Murad [1], in an earlier work looked at the equations Einstein used where he stressed the speed of light was a limitation. Einstein is absolutely correct within the parameters of his assumptions. However, if an assumption is relaxed where the observer moves faster than light speed, the equations indicate the relativity factor does not become imaginary but remains a real number for these faster than light conditions.

As soon as one mentions going faster than light speed, those that support the conventional wisdom usually and immediately have objections. It is very difficult to provide sound proof in the form of unambiguous evidence where such a possibility may exist. However, there are some anomalies in the cosmos that may provide such evidence. If the gravitational field of a black hole is so strong where even light cannot leave, jets arise and if they are not created or draw material from an accretion disk but from the black hole itself, then they must represent evidence of physical phenomenon moving faster than light speed. We are not saying this is true but rather a different way of looking at things is required to derive different insights. Thus the issue is to discover a jet from a black hole with no accretion disk to validate this rationale. Space travel a century ago was not a driving concern when Einstein [2] put forth his theories. In fact, discovering the electron was the major challenge paramount in that era as well as continuing to understanding how electric and magnetic fields interact with each other.

Some Russian scientists have a different approach to deal with faster than light phenomenon while still embracing Einstein's guidance. Dyatlov [3]-[7], a former Russian chief scientist of a mathematics institute located in Novosibirsk, Russia, was adamant about obeying Einstein's theory of relativity. However, like other Russian scientists such as Kosyrev, the Russians generally 'hedged' their bets regarding faster than light phenomena. For them, gravity would still move only at light speed; nevertheless, they might invoke a torsion field to support a propagation velocity possibly an order of magnitude faster than the speed of light. Godin [8] typifies these views and suggests the theory of torsion fields assumes a gravitational field corresponds to a longitudinal polarization of the physical vacuum. Being longitudinal waves in an elastic physical vacuum, the gravitational waves should have a very high speed of propagation, of the order of $10^{+9}c$. It's probable, the torsion fields with longitudinal polarization could be responsible for the long-range action and manifestation of Mach's principle.

So there are certain uncertainties that still exist. Thus our objectives are clear. If one really needs to develop a propulsion system capable of near light travel or faster, can the elements within Einstein's Field Equations be broken down to understand physical events that may or may not occur? Can these equations reveal the means to travel faster than the speed of light or identify potential hazards, which are currently speculative or even unknown?

II. DISCUSSION

Problems involving the theory of relativity mostly fall within the realm of physics. In general, physicists fiercely defend the tenets of General Relativity and Special Relativity* within the context of the conventional wisdom established for over a century. Unfortunately they do not build things where engineers do. Thus our desire is to explore the ramifications and potential where such things happen as a physicist suggests and are real but in a very general fashion that allows an engineer to also better participate within the creation process. Basically, the engineer also wants to understand these situations and subsequent consequences so a suitable propulsion scheme can eventually be devised.

As a given, the author makes no claim to being an expert in implementing or further developing these equations [1] other than a minor contribution. However, as an engineer, there is a necessity to understand these terms and the differences between the theory of General Relativity

---

* Basically, General Relativity treats detectors whose velocity vector need not be constant (accelerating) whereas Special Relativity has restrictions that such detectors are at constant velocities. The test particle velocities being measured are not restricted in either theory except they are not faster than light. The theory of special relativity can be derived formally from a small number of postulates:

1. Space and time form a 4-dimensional continuum where the word for a continuum implies a manifold.
2. The existence of globally inertial frames suggests there exist global spacetime frames with respect to unaccelerated objects moving in straight lines at constant velocity.
3. The speed of light c is a universal constant, the same in any inertial frame.
4. The principle of special relativity or the laws of physics are the same in any inertial frame, regardless of position or velocity.

and Special Relativity. One situation will not be discussed is when the tensors in the field equations may not be of the same order. This is a very difficult situation mathematically and physically to comprehend and is open to different interpretations. For example, one may argue the highest order tensor or the tensor with the most elements may represent effects from another dimension. This may directly influence living in dimensions. How these terms evolve is the subject of another treatment.

A. Einstein's Field Equations

Basically Einstein's field equations is part of a geometric theory, which examines the curvature of a space-time continuum based upon a gravitational field and various forces represented by:

$$R_{\mu\nu} - \frac{1}{2} g_{\mu\nu} R = \frac{8\pi G}{c^4} T_{\mu\nu} \tag{1}$$

This form does not include the cosmological constant, which accounts for an expansion or contraction of the universe. A more complete form would be:

$$R_{\mu\nu} - \frac{1}{2} g_{\mu\nu} R + g_{\mu\nu} \Lambda = \frac{8\pi G}{c^4} T_{\mu\nu} \tag{2}$$

where $R_{\mu\nu}$ is the Ricci curvature tensor, R the scalar space time curvature, $g_{\mu\nu}$ the gravitational metric tensor, Λ is the cosmological constant, G is Newton's gravitational constant, c the speed of light, and $T_{\mu\nu}$ the stress-energy tensor sometimes referred to as the mass-energy tensor or even the stress-energy-momentum tensor.

Einstein's field equations are a tensor equation relating a set of symmetric 4 x 4 tensors in its simplest form. It is written using abstract index notation. Each tensor has 10 independent components. Generally when fully written out, the field equations are a system of 10 coupled, nonlinear, hyperbolic-elliptic partial differential equations. Despite the simple appearance of the equations, they are quite complicated. Given the freedom of choice of the four space time coordinates, the independent equations reduce to 6 in number. Although Einstein's field equations were initially formulated in the context of a four-dimensional theory, some theorists have explored the consequences in n dimensions. Equations of higher dimension may be outside of general relativity but are still referred to as Einstein's field equations. The vacuum field equations define Einstein manifolds. Given a specified distribution of matter and energy in the form of a stress-energy tensor, Einstein's field equations are understood to be equations for the metric tensor $g_{\mu\nu}$, as both the Ricci tensor and scalar curvature depend on the metric in a complicated nonlinear manner.

B. The Cosmological Constant

Einstein introduced the term cosmological constant to account for a static universe (i.e., one that is not expanding or contracting). This effort was unsuccessful for two reasons: the static universe described by this theory was unstable, and observations of distant galaxies by Hubble a decade later confirmed our universe is not static but expanding. So Λ was abandoned with Einstein calling it the "biggest blunder [he] ever made". For many years the cosmological constant was almost universally considered to be zero.

Despite Einstein's misguided motivation for introducing the "cosmological constant" term, there is nothing inconsistent with the presence of such a term within the equations. Note the cosmological constant operates on the gravitational tensor. Indeed, recent improved astronomical techniques have found a positive value of Λ is required to explain some observations. Einstein thought of the cosmological constant as an independent parameter, but its term in the field equation, can also be moved algebraically to the other side, written as part of the stress-energy tensor or:

$$T_{\mu\nu}^{(vac)} = -\frac{\Lambda c^4}{8\pi G} g_{\mu\nu}. \qquad (3)$$

Another approach would be to include the term in the vacuum energy as a constant given by:

$$\rho_{(vac)} = \frac{\Lambda c^2}{8\pi G}. \qquad (4)$$

A positive cosmological constant is equivalent to the existence of a non-zero vacuum energy or demonstrates zero-point energy exists. The terms are now used interchangeably in general relativity.

C. The Stress-Energy Tensor

In Einstein's field equations, the stress-energy tensor is proportional to and represents a collection of conservation equations. In other words, each location in the matrix represents conservation of some quantity be it mass, momentum, energy, or some other relationship such as Maxwell's equations. Basically this mass-energy tensor, probably better named as a 'conservation' tensor, is normally zero or vanishes for a majority of realistic physical phenomenon. If, however, some of these equations violate conservation or produce excess not defined by conservation, the excess can create changes in the curvature of the space-time continuum or directly change the gravity tensor. Such events could occur with an over-unity process or during an anomalous event. It is difficult to say if these effects will influence one of these quotients more than the other or vice-versa. Basically one would like to enhance such effects in developing a propulsion system.

In practice, it is usually possible to simplify the problem by replacing the full set of equations of state with a simple approximation. Some common approximations by physicists are:

- Vacuum: $T_{\mu\nu} = 0$
- Perfect fluid:

$$T_{\mu\nu} = (\rho + p) u_\mu u_\nu + p g_{\mu\nu}$$

where $u^\mu u_\nu = -1$

- Non-interacting dust (a special case for assuming a perfect fluid): $T_{\mu\nu} = \rho u_\mu u_\nu$

Here ρ is the mass-energy density measured in a momentary co-moving frame, u_a is the fluid's 4-velocity vector field, and p is the pressure. These are non-dimensional quantities. For a perfect fluid, another equation of state relating density ρ and pressure p must be included. This equation will often depend on temperature, so either a heat transfer equation is required or the postulate is used where heat transfer can be neglected.

General relativity includes the conservation of energy and momentum expressed as:

$$\nabla_\nu T^{\mu\nu} = T^{\mu\nu}{}_{,\nu} = 0 \tag{5}$$

Again, if this last term does not vanish due to an anomalous occurrence, then we should expect changes to the other terms in the field equations. From Pharis Williams [10], the space energy-momentum tensor for matter under the influence of gauge fields is given by:

$$T^{ij} = \gamma u^i u^j + \frac{1}{c^2}\left[F^i_k F^{kj} + \frac{1}{4} a^{ij} F^{kl} F_{kl}\right] \tag{6}$$

And may be written in terms of the surface metric as:

$$T^{\alpha\beta} = \gamma u^\alpha u^\beta + \\ + \frac{1}{c^2}\left[F^\alpha_k F^{k\beta} + F^\alpha_4 F^{4\beta} + \frac{1}{4}(g^{\alpha\beta} - h^{\alpha\beta})(F^{\mu\nu} F_{\mu\nu} + F^{4\nu} F_{4\nu})\right]. \tag{7}$$

Using Maxwell's equation in a three dimensional Cartesian coordinate systems with time, we have:

$$F_{ij} = \begin{bmatrix} 0 & \dfrac{E_x}{c} & \dfrac{E_y}{c} & \dfrac{E_z}{c} \\ -\dfrac{E_x}{c} & 0 & B_z & -B_y \\ -\dfrac{E_y}{c} & -B_z & 0 & B_x \\ -\dfrac{E_z}{c} & B_y & -B_x & 0 \end{bmatrix}. \qquad (8)$$

The field equations of classical electromagnetism are Maxwell's equations which describe how electromagnetic fields are produced from charged particles and are written in the framework of special relativity devised to consistently describe electromagnetism and classical mechanics as:

$$F^{ij}{}_{,i} = k j^j \qquad (9)$$

This arises from the following Lagrangian:

$$L = -\dfrac{1}{4\mu_o} F^{ij} F_{ij} + j^i A_i. \qquad (10)$$

The variable j^i is a vector representing electrical and magnetic currents.

III. ANALYSIS

The material covered so far is basic from the conventional wisdom pertaining to the field equations. There is no controversy about the stress-energy tensor except for situations were non-conservation may occur. There is no reason to assess changes in the Cosmological Constant except if one may add another tensor to the problem indicative of a torsion field.

The following sections will address items in these equations that are altered or may lead to a different set of conclusions with propulsion implications. Some of these terms may have very strict constraints and the issue is to question the validity of assumptions as well as what may be required to remove these constraints. There are also situations where the conventional wisdom may not go far enough to provide explanations and possibly lead to a different interpretation. In other words, can the physics differ from the mathematics and can the mathematics provide different insights into the physics?

A. The Gravitational Tensor

Based upon recent events in the cosmos, one may argue Newtonian gravitation is inadequate for describing phenomenon such as the unusual high velocity of galactic spiral arms or even the Pioneer anomaly where there is a sudden increase in gravity in the direction of the sun at larger distances. It is claimed that to account for the observed deficiencies in the former situation, additional mass must exist. This has created the doctrine of 'Dark' matter[1] where matter is essential to make up for the shortcomings of Newtonian gravitation. To date, particles of dark matter, which are postulated for some arbitrary reason as being huge, have yet to be detected. One obvious solution to this problem is to accept the premise where perhaps, Newtonian gravitation has limitations and requires changes to account for these anomalies. Thus if there is an excess within the mass-energy tensor, it could feed directly into the gravitational terms and alter gravity or possibly the space-curvature tensor.

The gravitational tensor is of crucial interest especially if one is to build a warp drive. Usually physicists linearize gravitation by treating the gravitational tensor simply as a vector based upon the main diagonal elements in the matrix, because it simplifies the mathematical complexity. However, other gravitational laws exist; of these laws, they should asymptotically approach Newtonian gravitation as a limit where gravity disappears at infinity. Jefimenko [11] implies gravity is not only a means for attraction but also induces angular momentum. In other words, the first thing an uncontrollable satellite would do when inserted into orbit is to start spinning. Does the Earth's gravitational field induce angular momentum in the satellite? Moreover, each major planet rotates in the same direction and each sees the same face of a major moon when viewed from the major planet's surface. If there were inhabitants of other planets, they too would never see the far-side of these moons. Jefimenko claims this is due to angular momentum induced by gravity. He lays out four situations where a body passing another body creates 'serious' frame-dragging effects. Here angular momentum is conserved. All of these interactions involve rotation and cannot be explained using Newtonian gravitation. What this implies is the gravity tensor may have off-diagonal elements induce these effects and such elements indicate linearizing the gravity tensor ignores some of the

---

[1] There is another perspective about dark matter. During the big bang, we see in our view that all of these particles are 'positive' matter. By conservation, what occurs with 'negative' matter? Thus a rationale exists for dark matter.

physical realities of gravity. This could, for example, have non-symmetric terms to include off-diagonal elements in the gravity matrix. A spherical coordinate system would look like:

$$g_{ij} = \begin{bmatrix} g_{rr} & 0 & 0 & g_{rt} \\ 0 & r^2 & 0 & 0 \\ 0 & 0 & r^2 \sin^2 \theta & 0 \\ g_{tr} & 0 & 0 & g_{tt} \end{bmatrix} \quad (11)$$

Note the mathematical difficulties are enhanced if gravity depends upon time to generate gravity waves or the g_{rt}, g_{tr} terms do not vanish. If dealing with Newtonian gravitation, all these terms would vanish because time does not fit into Newton's gravitational equation. If other laws are considered, say Jefimenko's, then these terms are significant such as $g_{r\theta}$, $g_{\theta r}$. If these exist, the complexity of the problem increases and we may have to solve a complex wave equation to describe gravity waves within the gravitational field.

More importantly, the suggestion here is some of the off-diagonal terms may involve a non-zero value that also increases the nonlinearity of the field equations. This is the essential part of the problem. Murad [12] points out angular momentum can be converted into linear momentum. This also implies off-diagonal components may exist where effects from one degree of freedom, say angular momentum about an axis of rotation, may impact another degree of freedom, say linear momentum in a coordinate direction. For rigid bodies, the effects of off-diagonal elements resident within the inertia matrix can easily produce this effect through the conservation of angular momentum equations. We also see this with a bicycle, or a car where angular momentum is converted into linear momentum. There are also situations where coupling of these effects can influence the trajectory dynamics of a rigid body. Most pronounced is a 'Dutch' roll, which is a coupling with sideslip angle and roll momentum that causes an airplane or a train to move in one sideway direction and then to suddenly move in another or opposite direction in a cyclic sinusoidal fashion. Such effects result from off-diagonal terms in the moment of inertia matrix.

B. Binary Pulsars

Winterberg [12], [13] points out rotation of a large Magnetar, a star with a very strong magnetic field, could induce a repulsive gravitational source term. This was examined by Murad [14] and points out it is coincidental many binary pulsar systems containing a neutron star and a companion have masses that are almost identical[2]. If anything, the neutron star in these binary pulsars rotates at different rates from other binary stars and this rotation cancels out a portion of the inertial gravity source term. There appears to be a mutual process where the mass difference between the neutron star and its companion may actually dictate the neutron star's rotation rate so both celestial bodies within the binary system have comparable weights and could travel in the same elliptical or circular orbit. This notion tends to demonstrate Jefimenko's ideas as well.

C. Gas Dynamic Azimuthal Momentum Transfer

There is also some analytical work that investigates these possibilities within a gas dynamic plasma. Let us return to the issue of converting angular momentum into linear momentum. Karimov, Stenflo and Yu [15] examine such nonlinear coupling within radial, axial, and azimuthal flows within asymmetric cold plasma. The analysis indicates that despite flow asymmetries, energy in the radial and axial flow directions can be transferred into the azimuthal component degree of freedom but not vice versa. Moreover, flow oscillations did not accompany density oscillations. However, the model assumes the flow vorticity is aligned only in the axial direction, a restriction, which might have been responsible for the mentioned result. If we relax this condition, we can investigate how the violation of cylindrical flow symmetry affects the energy transfer among the different flow directions.

In another effort [16], they examine the same conditions but for a rotating flow producing similar results. Nonlinear electron velocity oscillations in the absence of electron density oscillations at the same frequency are shown to exist. The rotating plasma is one of the simple physical systems exhibiting unusual nonlinear wave and pattern phenomena. Finite rotating cold plasmas are of special interest for understanding plasma crystal formation and other phenomena.

Most investigations [16] of nonlinear wave fluid phenomena consider finite but small perturbations about an equilibrium or steady state condition. Thus, resulting nonlinear waves behave quite similarly to linear modes, and the problem is usually studied by carrying out small-amplitude expansions of the relevant physical quantities. Since many waves have similar dispersion, propagation, and nonlinear properties, their nonlinear behavior can usually be described. However, for systems far from equilibrium, the

---

[2] Obviously we are trying to answer solutions with minimal information. If the mas is the same, in a random cosmos, this has to be a coincident.

perturbative approach is not applicable. One must then use a full nonlinear treatment to investigate the possibility of steady state, waves, and patterns. Solutions describing an expanding plasma with oscillatory flowfields are then obtained. The solutions show the energy in the radial and axial flow components can be transferred to the rotating component but not vice versa.

Let us return to a specific situation in the stress-energy tensor. The result where energy in the azimuthal flow cannot be converted into the other degrees of freedom, but the latter can be converted into the azimuthal flow. Depending on initial conditions, oscillations can be limited to only the azimuthal flow component; can have important implications in applications involving rotating plasmas and fluids. There can last flow oscillations, not accompanied by density oscillations on the same time scale. However, the evolution is sensitive to initial conditions before the nonlinearity of the motion becomes predominant. Thus it is feasible to have none-symmetric stress-energy tensors.

Karimov and Godin [17] examined the interaction of radial and azimuthal nonlinear waves in a twirling, cold plasma cylinder. It demonstrated that energy exchange between radial and azimuthal modes can occur, which leads to the acceleration of the rotation of the plasma cylinder. They examined the processes of energy exchange in twirling two-dimensional (2D) plasmas, which may lead to acceleration of rotation for the whole plasma medium. The possibility of processes for twirling plasmas follows from the well-known phenomenon of energy localization of initial excitations in the chain of 1D nonlinear linked oscillators (so-called Fermi–Pasta–Ulam recurrence effect) takes place only under some initial conditions. Such a simple dynamic system, instead of manifesting the equipartition of energy to all degrees of freedom, showed an unbalanced distribution of energy. The nonlinearity and collective interaction of different modes is necessary for concentration of energy. One can also expect similar behavior for a 2D, time-dependent system. Owing to different kinds of interactions may exist in these systems due to their initial states, different modes of collective motion are possible. Therefore, unlike 1D systems, time-dependent nonequilibrium states for these systems will be more versatile in character and such a process in rotating plasmas can lead to energy transfer from the radial degree to the azimuthal degree.

If this assumption is correct, Karimov and Godin set out to study evolution of a neutral, collisionless plasma system initially in a state far from equilibrium. In a 2D rotating system, kinetic energy density can increase due to:

(i) The energy exchange between the oscillator modes (radial and azimuthal modes) and rotation motion and

(ii) The self-compression of the whole plasma cylinder coming from rotation of the plasma skeleton.

Focusing only on energy transfer and avoiding self-compression processes, they show the strong nonlinear interaction between different kinds of plasma waves in cylindrical geometry can lead to energy accumulation in the rotary mode, which is brought about by the energy transfer into the radial degree of freedom. They propose a pattern for a strong nonlinear, twirling 2D plasma system of finite size originally far from equilibrium. The approach demonstrated certain acceleration of the originally twirling plasma might be observed where energy production stems from redistributing the energy in space and time.

It is clear that the realization of such a mechanism requires some supply of 'internal' energy into the system such as an oscillator, electromagnetic form or other types including chaotic degrees of freedom. It is also worth noting evolution of a plasma system is not restricted by any physical mechanism, such as dissipative processes, which limits growth of the kinetic energy of the plasma cylinder. Nevertheless, an effect of the transformation of energy in nonlinear cylindrical plasma waves can occur in some real plasma systems. Thus, if these efforts are real, momentum and energy may be converted from one coordinate to another within a gas dynamic process.

D. The Curvature Tensor

Of all of the terms in the field equations, this tensor has the most ambiguity. Mathematicians have stepped forward to quickly define this term based upon the theory of relativity as a geometric theory. Elements in this tensor may mean more than what is suggested by the conventional wisdom. If not, then there is obviously a need to extend the field equations further to consider these implied effects.

In differential geometry, the Ricci curvature tensor, named after Gregorio Ricci-Curbastro, represents the amount by where the volume element of a geodesic ball in a curved Riemannian manifold deviates from that of the standard ball in Euclidean space. It provides a way of measuring the degree to where the geometry is determined by a given Riemannian metric that might differ from a metric in ordinary Euclidean n-space. More generally, the Ricci tensor is defined on any pseudo-Riemannian manifold. Like the metric itself, the Ricci tensor is a symmetric bilinear form on the tangent space of the manifold.

The Ricci curvature is broadly applicable to modern Riemannian geometry and general relativity theory. In connection with the latter, it is up to an overall trace term, the portion of the Einstein field equation representing the geometry of space time, the other significant portion of which comes from the presence of matter and energy. Lower bounds on the

Ricci tensor on a Riemannian manifold allow one to extract global geometric and topological information by comparison with the geometry of a constant curvature space form. If the Ricci tensor satisfies the vacuum Einstein equation, then the manifold is an Einstein manifold. In this connection, the Ricci flow equation governs the evolution of a given metric to an Einstein metric. As a consequence of the Bianchi identities, the Ricci tensor of a Riemannian manifold is symmetric in the sense that: $Ric(\xi,\eta) = Ric(\eta,\xi)$. This is true, more generally, if the Ricci tensor is associated to any torsion-free affine connection for which there exists a parallel volume form. It thus follows the Ricci tensor is completely determined by knowing the quantity $Ric(\xi,\eta)$ for all vectors ξ of unit length. This function on the set of unit tangent vectors is often simply called the Ricci curvature, since knowing it is equivalent to the Ricci curvature tensor. It is also convenient to regard the Ricci tensor as a symmetric bilinear form. To this end for vector-fields X, Y, we can write:

$$Ric(X,Y) = X^i Y^j R_{i,j}. \tag{12}$$

The question is what occurs if we are not dealing with Reimannian geometry? What if there is no symmetry and if the main diagonal elements define required conditions? You really need a Ricci vector and not a tensor. However, what is the significance of off-diagonal elements?

Lack of symmetry leads to the following equation where this creates additional terms. These terms can influence other terms within the Field Equations to include gravity. Formally, this means:

$$\frac{\partial^2 T_r}{\partial u \partial v} - \frac{\partial^2 T_r}{\partial v \partial u} = R_{rpmn} T^p \frac{\partial x^m}{\partial u} \frac{\partial x^n}{\partial v}, \tag{13}$$

Where T_r is any covariant vector and:

$$T^p = a^{pq} T_q. \tag{14}$$

The RHS would feed directly into creating unexpected gravitational effects.

E. A Gravitational Distortion

Basically the Ricci tensor represents the curvature in the space-time continuum. If curvature exists, then it implies the existence of a field, usually a gravitational field. Other fields such as an electric, magnetic or even a torsion field could endure to create curvature; however, the field strength would have to be considerable. If the mass-energy tensor identically vanishes but both the gravitational and space curvature tensors remain, they can directly influence or drive each other. One may possibly call this situation a distortion within the space-time continuum or:

$$R_{\mu\nu} - \frac{1}{2} g_{\mu\nu} R = 0 \tag{15}$$

This is sometimes referred to as the vacuum field equations. For example, black holes in several galaxies must be spinning at or near their maximum rates. Per Rodrigo Nemmen [17] where:

"We think these monster black holes are spinning close to the limit set by Einstein's theory of relativity, which means that they can drag material around them at close to the speed of light."

Such a possibility may yield a space-time continuum distortion to satisfy these conditions. This suggests an off-diagonal element in the gravity tensor can induce unusual effects within the Ricci curvature tensor. This could also include the causal effects of a gravitational vortex, which may be created as a consequence of Gertsenshtein's ideas relating gravity to both electric and magnetic fields.

If one examines the elements of this Ricci tensor within accepted convention, it is easy to see if the gravitational tensor were vectorized, the Ricci tensor would be as well. In other words, you only really need elements on the main diagonal to uniquely define the curvature in your specific space-time continuum. However, if an off-diagonal element occurs in the gravitational tensor, it is most likely that the same situation will exist here. The notion is each matrix location may imply a different coordinate within a multi-dimensional space-time continuum. Does this imply the ability to travel in parallel universes by jumping from one matrix location to another within the Field Equations? What is the significance of off-diagonal elements in this tensor? If anything, the matrix could represent a topological manifold to lead toward these different dimensions and universes or multi-verses. Could this also include coordinates where you live under the influence of exponential time? The question here is if one can live in such a continuum where both linear time and exponential time can co-exist? If the significance of only diagonal elements implies the realm of a linear world with 'real' dimensions, then does the presence of off-diagonal elements represent a nonlinearity that implies of going from one space-time to another?

If we travel at the speed of light, the desire may transit from point A to point B. However, if there is an off-diagonal element in this tensor, does it mean we can travel to a parallel universe moving along a parallel coordinate axis, or even go back and forth within time? Clearly the want to understand the meaning of these off-diagonal elements within a physical context is required. If important, then it behooves our space drive propulsor to create and amplify these terms if we are to travel at the speed of light or go into a space-time continuum that can represent both the past and/or the future.

F. Ricci Scalar Curvature

This is a very difficult term to define and yet it provides physical significance to clearly understand. Einstein [2] defines this term as: "the scalar curvature formed by repeated contraction". Einstein goes on to state: "the condition $R = 0$, which would have to hold good everywhere for the $g_{\mu\nu}$, independently of the electric field."

This variable has a significant role to play; however, its value may be arbitrary. For a zero value of the scalar curvature, there is no correlation between the stress-energy tensor and the gravity tensor. If anything, any effect from the stress-energy tensor will only influence the curvature tensor. If this changes, then the scalar curvature will change. Thus curvature will now influence the gravity tensor.

Ricci's variable may be useful in describing space time singularities. Amongst singularities, there are two important types of space time singularities which are *curvature singularities* and *conical singularities*. Singularities can also be divided according to whether they are covered by an event horizon or not referred to as a naked singularity. According to general relativity, the initial state of the universe, at the beginning of the Big Bang, can be considered as a singularity. Another type of singularity predicted by general relativity occurs inside a black hole: any star collapsing beyond a certain point would form a black hole, inside which a singularity exists, as all the matter would flow into a certain point (or a circular line, if the black hole is rotating). These singularities are also known as curvature singularities. Here, a scalar curvature that is infinite in space time can describe a singularity in general relativity. We note these singularities are often associated with an incomplete description of geodesics. That is where geodesics cannot be smoothly extended due to unbounded curvature on their incomplete ends.

IV. RESULTS

The major point is unusual conditions within the field equations. These include where tensors are not vectorized and need to be treated as a complex wave partial differential tensor equation. Situations were discussed where off-diagonal elements can exist in either the gravitational or Ricci tensor or what would be significant. These clearly have propulsion ramifications. The basic problem is now one of how to make these things occur from an engineering discipline perspective. This is not trivial.

A. Over-Unity Situations

There is an interesting possibility, which should not be ignored and these are specific conditions where there are imbalances in equilibrium or situations where you can create more energy than what exists in a control volume. A nuclear explosion is a typical example and subsequent EMP effects occur on very small time scales would demonstrate this. However, nuclear explosions are usually not desired for a propulsion system because they create structural and thermal problems and influence an astronaut's biological properties on board the craft.

Creating an *excess* within the stress-energy tensor may be important. One approach suggested by Barrett [19], [20] involves looking at Maxwell's equations where there is a more detailed treatment of boundary conditions. This also assumes a different gauge is used that suggests some of the mathematical operations may not commute or there are problems with associativity. Barrett[3] goes one step further by using these conditions to identify a new electromagnetic field or an 'A' field based upon these conditions shown in Tables 1 and 2.

When included within the field equation, the stress-energy tensor will have additional terms to feed into both the Ricci tensor and Gravitational tensor. This will create space-time curvature. As previously mentioned, the same would hold for a propulsion system could create off-diagonal elements in these two tensors as well.

Other conditions would increase the vector that contains currents for these equations. For example, the influence of a magnetic current could also be in the direction in our favor as well. This could also include creation of magnetic monopoles.

B. Issues Related to Einstein's Field Equations

Einstein's field equations are a system of 10 coupled, nonlinear, hyperbolic-elliptic partial differential equations. Despite the simple appearance of the equations, they are quite complicated. Given the freedom

---

[3] Regarding Barrett's work, the purist may not accept his work by virtue of using different vector/scalar terminology. The problem here is to understand his genius in doing this rather than conventional approaches.

of choice of the four space time coordinates, the independent equations reduce to 6 in number. Although Einstein's field equations were initially formulated in the context of a four-dimensional theory, some theorists have explored the consequences in n dimensions. Equations of higher dimension may be outside of general relativity but are still referred to as Einstein's field equations. The vacuum field equations define Einstein manifolds. Given a specified distribution of matter and energy in the form of a stress-energy tensor, Einstein's field equations are understood to be equations for the metric tensor $g_{\mu\nu}$, as both the Ricci tensor and scalar curvature depend on the metric in a complicated nonlinear manner.

Table 1. A New Way of Viewing Maxwell's Equations [19], [20]

	$U(1)$ Symmetry Form (Traditional Maxwell Equations)	$SU(2)/Z_2$ Symmetry Form
Gauss' Law	$\nabla \bullet E = J_0$	$\nabla \bullet E = J_0 - iq(A \bullet E - E \bullet A)$
Ampères Law	$\dfrac{\partial E}{\partial t} - \nabla \times B - J = 0$	$\dfrac{\partial E}{\partial t} - \nabla \times B - J + iq[A_0, E]$ $- iq(A \times B - B \times A) = 0$
	$\nabla \bullet B = 0$	$\nabla \bullet B + iq(A \bullet B - B \bullet A) = 0$
Faraday's Law	$\nabla \times E + \dfrac{\partial B}{\partial t} = 0$	$\nabla \times E + \dfrac{\partial B}{\partial t} + iq[A_0, B] =$ $iq(A \times E - E \times A) = 0.$

Thus Einstein's field equations relate the differences between the stress-energy tensor, the gravity tensor, and curvature tensor. As mentioned, physicists do not generally accept the way engineers use a constant value for gravity. They use a gravity tensor instead a constant and assume there are only terms on the main diagonal of the gravity tensor and all the off-diagonal terms go to zero. Interestingly, this now goes into the same term as the engineer but after considerable mathematics. In other words, the physicist's main diagonal terms are the engineer's gravitational constant value. The only problem is if this metric tensor off-diagonal term is not zero in nature, there will be some strange behavior and we have yet to thoroughly investigate or experimentally understand this.

Table 2. The Impact Upon More Detailed Specification of Boundary and Other Conditions [19], [20]	
$U(1)$ Symmetry Form (Traditional Maxwell Theory)	$SU(2)/Z_2$ Symmetry Form
$\rho_e = J_0$	$\rho_e = J_0 - iq(A \bullet E - E \bullet A) = J_0 + qJ_z$
$\rho_m = 0$	$\rho_m = -iq(A \bullet B - B \bullet A) = -iqJ_y$
$g_e = J$	$g_e = iq[A_o, B] - iq(A \times B - B \times A) + J =$ $= iq[A_o, E] - iqJ_x + J.$
$g_m = 0$	$g_m = iq[A_o, B] - iq(A \times E - E \times A) =$ $= iq[A_o, B] - iqJ_z.$
$\sigma = J/E$	$\sigma = \dfrac{\{iq[A_o, E] - iq(A \times B - B \times A) + J\}}{E} =$ $= \dfrac{\{iq[A_o, E] - iqJ_x + J\}}{E}.$
$s = 0$	$s = \dfrac{\{iq[A_o, B] - iq(A \times E - E \times A)\}}{B} =$ $= \dfrac{\{iq[A_o, B] - iqJ_z\}}{H}.$

For example, rotation may create linear momentum can be considered as nonsense. However, we do it all the time on using a car, bicycle or a train. If the off-diagonal terms in the inertia tensor are large, the impact of Dutch-roll is more exaggerated. Obviously if an off-diagonal term exists in the gravity tensor, this would have space propulsion implications as well. However, the physicists or anyone else does not work on this problem because the mathematics is too difficult. To the engineer, these are still realistic considerations that need to be simulated.

The stress-energy tensor really represents the sum of all conservation equations as well as Maxwell's equations that engineers know, believe and accept. We engineers live and die by conservation. Aerodynamicists trace streamlines, electrical engineers trace electrons, nuclear engineers trace neutrons and so on.

If we have situations violate conservation, engineers really do not know what to do. Unknowingly the physicists have a solution with the Field Equations. Here, if there is some excess in conservation, it spills over into changing the gravitation tensor or the curvature tensor. If conservation is satisfied, then the stress-energy tensor goes to zero and these two terms, the curvature tensor and the gravitational tensor, interact with each other where space-time curvature is now a direct function of the gravity tensor and vice-versa. This results in:

$$R_{\mu\nu} - \frac{1}{2} g_{\mu\nu} R = 0 \tag{2}$$

The problem is by conventional wisdom, this violates the rule to validate conservation to define if a concept is either fake or real. In other words if you look at a concept and it violates conservation, it may be a fake idea or it may be the consequences of legitimate gravitational or other types of effects. Guess what, this is not heresy although it goes against the conventional wisdom. This means one has to be very careful when they see these effects and it might warrant more thorough examinations.

In many situations, engineers and scientists tend to linearize complex systems in the hope of creating a solution to a particular problem. By walking away from the more difficult challenges posed by a nonlinear problem, they lose sight of potential solutions to solve real problems of interest. Linearizing the field equations by vectorizing the gravitational tensor into a vector to satisfy Newtonian gravity requirements creates a vectorized Ricci tensor as well. Problems generally with the stress-energy tensor are straightforward but now we are solving a greatly simplified problem provide no insight to the initial more interesting solutions.

Let us look at this further. The physicist claims singularities exist in the metric for defining the field equations. The metric is used to derive a space warp drive or to build a wormhole. However, as an engineer, we have no idea how to create such singularities in the metric. Others claim a singularity must satisfy conservation and physically, this is hard to fathom let alone to examine or calculate the value of a singularity. What do you do? Measure a line integral like a complex variable surrounding the singularity? Maybe this works but only in an analytical context. However, what about physical realities? How do we measure conservation using magnetic and electrical fields or torsion fields?

Let us look at several important issues related to conservation. We talk about the Dirac Sea. What created the first time that particles to induce the Dirac Sea were instantaneously created for the first time? What energy level was required to achieve this to occur? What about the current vibration of the cycle of these particles? Did they do this with non-conservation? Was this God?

The Dirac Sea occurs where particles obey a cycle of life and death also to change their electric, magnetic, and gravitic fields during this cycle. The question is when this occurs, is their one particle? How about 100 particles? Is it a million or a hundred million billion particles per cubic inch? We never really discuss talking about the density of the ZPF. For example, between galaxies, we can assume the number is very low because there would be gravitational attractions to attract these free particles to nearby galaxies even if the attraction is but a small instance. This would be a real vacuum between galaxies. Unfortunately, we do not know enough about the ZPF [21] let alone have the ability to assess its density or intensity may be worthwhile exploiting for creating 'free' energy.

If we look at a theory about the Big Bang, one view is that everything may have spilled out of the fifth dimension in a tremendous amount to fill in our four-dimensional space. How do we conserve the fifth dimension? How valid is Kalaza-Klein [22] to conserve a fifth dimension?

People ask for other dimensions. If they talk about metrics, why do they not include these terms to define new dimensions within the metrics? Moreover, this could also include multiverses as well. Maybe the Big Bang really does not work? Maybe it is real!

No one really seems to explain a theory to allow us to go from one dimension to another. Let us offer a hypothesis. We do not talk much about the curvature tensor. It is most likely filled within the main diagonal term and the off-diagonal terms vanish. If we extend the field equations to increase other dimensions such as in a multiverse, this tensor size will increase. If an off-diagonal term appears in this tensor, then it allows us to go from one dimension to another. The obvious question is how to make such an off-diagonal term to exist and what is the mechanism to establishing experimental verification to go from one dimension to another? Remember the point [23] is:

Truth, some say, is what agrees with experiment.

This point is worth suggesting if engineers and physicists have to accept over unity items as being reality and not magic. People may really be making things interact with the ZPF or the Dirac Sea but we need to be more careful to understand and try to calibrate how to do this. If we do not, we may offer opportunities cancel some real physical events that could take us to the sky. The other point is due to our intellectual limitations, our trend to linearized nonlinear problems as well as ignoring off-diagonal terms on matrices or tensors, may create our ignorance and actually prevent us from discovering the truth.

We have many serious problems to understand about the theory of relativity. For example [23]:

No length contraction has ever been shown on a well-defined, charged or uncharged body with well-defined dimensions and a velocity measured by several independent methods, if not directly; no time dilation experiment has ever provided proof that the charged rate of a clock is only perceived by the moving observer and has not taken place in the clock itself.

The Einstein theory has never proved its two tacit postulates: that the Maxwell-Lorentz electrodynamics remain valid at high observer-referred velocities; and that the motion of matter through a force field does not inherently- independently of any observer- change its own field.

Some individuals feel the theory of relativity is well-grounded with considerable experimental evidence. This may be the case. However, to improve our capabilities, we want to always question the conventional wisdom. What would be a valid experiment to demonstrate time-and spatial dilation? From the same reference:

Let us take the opportunity to dispel another myth, namely that Einstein's theory contradicts Newton's laws. The statement that force equals mass times acceleration was put in Newton's mouth posthumously; there is no place in the Principia where Newton makes such a statement. He always writes about the rate of change of momentum (mutatio motus, or "change of motion,"). In present notation – the Principia makes their case by geometry – Newton never took the m out of the parentheses in d (mv)/dt, for he was too careful a man to ignore the possibility that internal mass might be a variable. When Einstein introduced velocity-dependent mass explicitly, he did not have to change one iota in Newton's Laws of Motion for any part of his theory.

This implies there is some commonality where both laws are supportive. What is important is Newton did not break-up the momentum to treat mass and acceleration. To some degree, Newton may have assumed there was some sort of interaction where the mass would not be a constant but variable as in the case of a rocket during its phase of burning propellant.

We deal with many constants used for many of our capabilities within physics or engineering. The question is we have to assume these effects are biased because of the reference of the laboratory or the Earth. This reference raises an interesting point:

The alternative to the Second Postulate that I will work with is the velocity of light is constant with respect to the local gravitational field through which it propagates.

Thus in all optical experiments supporting the Einstein theory, the observer was always nailed to the gravitational field of the Earth.

What does this mean? Simply viewed, we only see things based upon our laboratory on the Earth to represent a reference system. However, gravity or the motion of the planet within the solar system may create a bias to impact what we observe on the Earth. For example, we will discuss later about a specific gravitational law with distance as well as velocity to define gravitation. What this suggests is when we go to space in the far-aboard, we should expect to see changes and our understanding may be altered. For instance, we look at the Doppler shift and low and behold, the universe is expanding in all directions. The problem is these signals may go by large gravitational sources to alter the waveform and provide wrong results. We should expect all sorts of changes.

Another point is the question of the cosmological constant. Originally this was to determine contraction or expansion of the cosmos. Later on it was used for cold energy and now is related to dark energy. The problem is the field equations are not carved in stone but represent a living capability. The cosmological constant probably represents a 'Fudge' factor to treat unusual phenomena. Furthermore, we should see additional terms in the field equations to consider Yilmaz's treatment to deal with many-body problems and mostly, the cosmological constant may really be a variable once to use a different reference laboratory.

V. SOME CONCLUDING THOUGHTS

Examining the field equations has provided an engineering framework to look at developing a far-term space drive. It reveals a situation could lead to a gravitational distortion, which may also have propulsion implications especially in the realm of creating a wormhole mouth or a means of traversing across the cosmos. With this possibility as well as creating off-diagonal elements in these tensors, the challenge is now up to the engineering disciplines to develop the technology that creates these anomalous conditions within a propulsor and uses them to explore and colonize both the near and far abroad.

These activities require evaluation within Einstein's field equations from an engineering perspective to possibly create a suitable propulsion system. A set of circumstances was uncovered where it may lead to an anomalous gravitational distortion in what is normally referred to as the "vacuum field equation" in a near flat space-time continuum with no electric, magnetic or torsion fields. The distortion, possibly created by a gravitational vortex, is driven purely by the gravitational tensor. Additionally these field equations imply unusual consequences may follow if one includes nonlinearities, off-diagonal terms in the gravitational or space-continuum tensors, more detailed boundary conditions as well as creating systems to enjoy over-unity performance. Depending upon definition of the curvature tensor, conditions for travel in a parallel universe, faster than light travel and past/future time travel may exist. These capabilities should be incorporated within a design of a suitable propulsor to colonize and explore unknowns within the cosmos.

Acknowledgments

The author wishes to thank assistance from Mr. Dana Sweet regarding the original manuscript as well as Dr. Jack Sarfatti for some suggestions. I would also like to thank Dr. Terrence Barrett for the tables considering changes in treating Maxwell's equations and Dr. Ivan Kruglak for the Russian references.

References

[1] Murad, P. A.: "Faster Than Light Travel Versus Einstein", Vol. 13, No. 3, *Galilean Electrodynamics*, May/June 2002.
[2] Lorentz, H. A., Weyl, H., and Minkowski, H.: *Einstein: The Principle of Relativity*, Dover Publications, Inc. 1952.
[3] Dyatlov, V. L.: *Electrogravimetric Energy Transformation*, Moscow, NT-Center, 1995, p. 29.

[4] Murad, P. A., Dyatlov, V. L., and Dmitriev, A. N.: "Interesting Problems of the Inhomogeneous Physical Vacuum", *Proceedings of the Science 2000 Congress*, St. Petersburg, Russia, July 3-8, 2000.

[5] Murad, P. A., Dyatlov, V. L., "Comparing the Inhomogeneous Physical Vacuum and the Zero-Point Field", *Proceedings of the Science 2000 Congress*, St. Petersburg, Russia, July 3-8, 2000.

[6] Lavrentiev, M. M., Dyatlov, V. L., Fadeev, S. I., Kostova, N. E., and Murad, P. A.: "Rotation Effects of Bodies In Celestial Mechanics", *Proceedings of the Science 2000 Congress*, St. Petersburg, Russia, July 3-8, 2000. Co-authors are Lavrentiev, M. M., Dyatlov, V. L., Fadeev, S. I., and Kostova, N. E. from Novosibirsk, Russia.

[7] Murad, P. A. and Dyatlov, V. L.: "An Ansatz on the Possibility of Hyper-Light Travel", *Proceedings of the Science 2000 Congress*, St. Petersburg, Russia, July 3-8, 2000.

[8] Kruglak, I.: Private conversation 2009.

[9] Information is extracted from Wikipedia, 2010.

[10] Williams, Pharis E.: *Mechanical Entropy and Its Implications*, Entropy 2001, 3, 76-115, ISSN 1099-4300, published 30 June 2001.

[11] Jefimenko, O. D., *Gravitation and Cogravitation*, Electret Scientific Company Star City, ISBN 0-917406-15-X, 2006.

[12] Murad, P. A.: "An Anzatz about Gravity, Cosmology, and the Pioneer Anomaly", presented at the February 2010 *SPESIF Meeting* at John Hopkins University.

[13] Murad, P. A.: "The Challenges of Developing the Technology for A Realistic Starship Propulsor", AIAA Paper 2010-1609 presented at the 48th AIAA Aerospace Sciences Meeting, January 4-7, 2010 in Orlando, Florida.

[14] Murad, P. A.: "An Alternative Explanation of the Binary Pulsar PSR 1913+16", presented at SPESIF, February 2009 in Huntsville, Ala.

[15] Karimov, A. R., Stenflo, L., and Yu, M. Y.: Coupled flows and oscillations in asymmetric rotating plasmas, *PHYSICS OF PLASMAS* 16, 102303, 2009.

[16] Karimov, A. R., Stenflo, L., and Yu, M. Y.: Coupled azimuthal and radial flows and oscillations in a rotating plasma, *PHYSICS OF PLASMAS* 16, 062313, 2009, published online 29 June 2009.

[17] Karimov, A. R. and Godin, S. M.: Coupled radial–azimuthal oscillations in twirling cylindrical plasmas, *IOP PUBLISHING PHYSICA SCRIPTA*, Phys. Scr. 80 (2009) 035503 (6pp) doi: 10.1088/0031-8949/80/03/035503, Published 25 August 2009.

[18] "Rapidly spinning black holes might be source of jets that affect galaxy growth", January 11th, 2008 - 3:00 pm ICT by admin Washington, Jan 11 (ANI).

[19] Barrett, T.W., *Electromagnetic phenomena not explained by Maxwell's equations. In Essays on the Formal Aspects of Electromagnetic Theory,* A. Lakhtakia (ed), World Scientific Publishing Co., 1993.
[20] Barrett, T.W., *Topological Foundations of Electromagnetism*, World Scientific, 2008.
[21] Takaaki Musha: Physics of the Zero Point Field and Its Applications to Advanced Technology, published September 1st 2012 by Nova Science Publishers.
[22] Duff, M. J.: "Kaluza–Klein Theory in Perspective". In Lindström, Ulf (ed.). *Proceedings of the Symposium 'The Oskar Klein Centenary'*. Singapore: World Scientific. pp. 22–35. ISBN 981-02-2332-3, (1994).
[23] Petr Beckmann: Einstein Plus Two, The Golem Press, Boulder, Colorado, 1987.

Nomenclature

B Magnetic field intensity (volt-second/m$^{2)}$
c Speed of light (3×10^8 meters/sec.)
E Electric field intensity (volts/meter)
g Gravity (9.8 m/sec^2)
J Current (coulomb/sec)
m Mass (kilograms)
p Pressure
R Ricci curvature tensor
T Stress-energy tensor

Greek Symbols

ε permitivity (farads/meter)
μ Permeability (henrys/meter)
σ Conductivity (mhos/meter)
Λ Cosmological constant
γ Relativity factor
ρ Volume charge density (coulombs/m^3)

Chapter II

Pulsar Behaviour that may impact a Future Space Propulsor

Abstract: Binary pulsars demonstrate unusual gravitational behaviour that might be a careful balance between orbit performance, companion star and neutron star weights, and most importantly neutron star spin rate. Some binary pulsars are believed to have the same weight for both stars moving in a highly elliptical orbit while some binaries with supposedly vast weight differences, are in near-circular orbits. This is counterintuitive and raises questions outside of the conventional wisdom. Moreover, each neutron star in these binary systems spin at different rates, which implies rotation per Winterberg's conjecture may induce a repulsive gravitational source. This is analogous to generating dark matter thereby negate inertial effects to allow balance. Swirling jets leaving black holes may move either at greater than light speed or also become a natural repulsive gravitational source. Interactions between the pulsar binary's two bodies and behaviour of two of Jupiter's moons may also validate notions from Jefimenko, who claims gravity induces linear attraction and angular momentum. If true, these findings show a significant relationship, which may exist between gravity and angular momentum as well as suggest angular momentum can be converted to linear momentum after going through an intermediate step of repulsive gravitation. Furthermore, these findings may be the prerequisites to devise future starship propulsion drives to explore the cosmos.

PACS: 04.50.Kd, 04.80.Cc, 06.30.Dr, 06.30. Gv, 97.10.-q, 97.10.Gz, 97.10.Xq, 97.60.Gb, 97.60.Jd, 97.60.Lf, 97.80.-d.
Keywords: Binary Pulsar, Neutron Star, Asteroid, Gravity, Jefimenko, Rotation, Angular Momentum, Trajectories.

I. INTRODUCTION

One wonders how discover and revealing nature's secrets to provide valuable insights for developing an advanced propulsion system. This is an ongoing investigation [1], [2] to examine the premise that angular momentum may be converted into linear momentum with the notion of eventually creating a propulsive drive. Basically, this premise may initially appear outrageous but a car, a bicycle, locomotive or a propeller-driven airplane essentially converts angular momentum into linear momentum. Ground transportation capability does so by using the ground to achieve traction while the airplane goes through an intermediate process where the propeller induces a force by increasing the linear momentum of the surrounding air to create forward motion. What would be the mechanism or intermediate step for creating such traction in space viable for this to occur in a space ship drive?

In a previous paper [1], the author examined this idea concerning angular momentum by exploiting anomalous behaviour associated with both neutron stars and black holes. Jeong [3] suggests jets leave both types of celestial bodies. The problem may be straightforward for a neutron star because there are no physical constraints regarding the speed of light. However, the issue for a black hole is far more complicated. Murad [1], [4], and [5] discussed the gravitational anomalies involving aspects of angular momentum that Jefimenko identified could not be reproduced by Newtonian gravitation. The latter suggests gravity is only an attractive force. Moreover, Jefimenko [6] created his own gravitational model to include both a gravitational field and a weaker cogravitational field. He claims gravity is not only an attractive force but also induces angular momentum. At relativistic speeds, Jefimenko suggests the gravitational field acts like an electric field while the cogravitational field acts, based upon Heaviside's notions, like magnetism. These effects alter the law where gravity is not only a function of distance but can be a function of both distance and velocity. Binary pulsars were examined to see if they could offer any similar insights coupling this notion where gravity could be an attractive force and source of angular momentum with the implication rotation could also induce a repulsive gravitational source. It is suggested in Murad [1], [4], [5], and [7] where older and more established binary pulsars, which normally consist of a neutron star and a companion, usually move in a single elliptical trajectory.

Let us provide some information regarding pulsars [8]. Pulsars are the original gamma-ray astronomical point source. Gamma ray telescopes detected this during the sixties. By 2010, there were 1800 pulsars based upon radio detections but only about 70 from Gamma ray telescopes. X-ray pulsars also exist. Radio, optical, X-ray or Gamma rays are often referred to as 'spin powered pulsars' assumed to be derived principally from the sources of energy coming from the neutron star's rotation. This radiation from a neutron star such as Gamma rays- electromagnetic radiation, are not massive particles, but of high frequency and high energy per photon. X-rays are electromagnetic radiation at shorter frequencies than the UV range and longer than gamma rays. Distinctions are not obvious for X-rays and gamma rays. Electrons emit X-rays where gamma rays are emitted by the atomic nucleus. These jets of particles that leave the neutron star are moving almost at the speed of light.

Binary pulsars consist of a neutron star, which move either in the same or separate trajectories with a companion. In general the neutron star weight of about 1.4 times the weight of our sun. This companion can consist of another neutron star, a typical star or a benign body. These rotate continually about each other and in some cases, energy is consumed in the form of radiation might result in unusual trajectory behavior to be discussed later.

For this Keplerian motion to occur, analysis implies both bodies in the binary must be of comparable weight. This appears to be coincidental where both bodies just happen to possess the same weight fraction. How can this occur if the celestial bodies moving throughout the cosmos have a random mass distribution? Something more fundamental is at work here! Some binary pulsars do not meet these requirements and their trajectory parameters degenerate to collapse two separate elliptical orbits into a single orbit. It is suggested the loss in the pulsar's trajectory energy results in generating a large spectrum of energy in the form of gravitational waves. However, without a gravity wave detector, it is difficult to experimentally characterize such waves, their waveform, or their frequencies except only theoretically.

II. OBJECTIVES

The objective is to find several binary pulsars to examine their weight distribution; specifically if the neutron star and companion have similar weights. However, if the neutron stars have significantly different rotation rates, this implies different repulsive gravitation source strengths which may exist in each of these binaries. Murad [1] provides a formula indicating a repulsive gravitational source may cancel some of the inertial gravitational source thereby making both the neutron star and its companion of what may appear, comparable weight. This supports Jefimenko's conjecture where gravity is also attractive compared with Newtonian gravitation, and may create a source for angular momentum. The intention is to examine this possibility further and evaluate any field propulsion implications.

III. METHODS

There are several issues regarding mechanisms or physical phenomenon with propulsion implications. Three separate issues will be discussed to include:
- Conversion of angular momentum into linear momentum,
- The possibility of Black Hole jets creating repulsive gravitation, and
- Dynamics for the formation of pulsars.

Remember foremost that the overall objective is to uncover propulsion implications, which will allow mankind for exploring the near- and the far-abroad. Their interrelationships will be summarized in the Conclusions.

A. Angular Momentum Conversion

The author implies in a previous paper [2] where angular momentum could be transferred into linear momentum especially if nonlinear effects are realized; say for example, in Einstein's field equations using an asymmetric gravity tensor. The gravity tensor is usually evaluated to be linearized or a gravity vector. This means that only main diagonal terms exist and off-diagonal terms are ignored. If an off-diagonal element does last, effects from one conservation equation would stream into another and the same would hold for different space-time continuums. The latter effects would be difficult to discern especially since we only live in a four dimensional space-time with assumed linearized spatial and temporal dimensions. It would be meaningless and beyond our current intellectual grasp to relate these to appear unseen with many more dimensions that span an expanded space-time continuum.

In fact if a space-time could endure possessing an additional dimension to include linear time while a fifth dimension includes exponential time. It is human nature to assume a fifth dimension can be characterized as an additional spatial or physical dimension. Such events for five dimensions are hard to visualize or sense. In such a continuum, you could tell time using both a linear and an exponential timepiece which correctly coincides with the same time. In fact we can do this in our current space-time continuum where memory exponentially reaches a past time if we desire to recall and bring these events to clear focus in the present. Unfortunately only our imagination could predict a future event in our world but such a prediction would be fraught with huge uncertainties.

B. Jefimenko's Conjecture

Jefimenko's notions (*e.g.* [7]) are contrary to Newtonian gravitation, which is considered only as an attractive force. Is there any semblance of proof where this is true especially when there is either none or very little evidence for an experiment?

Two moons of Jupiter, Himalia and Elara are probably more recently captured asteroids, which have not yet had sufficient time to synchronize their orbits, periods and rotational rates. Himalia is the tenth known satellite of Jupiter. As the brightest of Jupiter's outer satellites, the Cassini mission captured Himalia for the first time. This was in a series of narrow angle images taken on December 19, 2000 from a distance of 4.4 million kilometers during a brief period when thrusters instead of reaction wheels stabilized Cassini's attitude. It is likely Himalia is not spherical but is believed to be an irregularly shaped asteroid.

Results indicate Himalia is at a distance of 11,480,000 km from Jupiter's surface and rotates every 0.4 days but requires 250.6 days to complete a revolution around Jupiter. So, an observer on Jupiter would certainly be exposed to all sides of Himalia, and similarly with Elara. Very little is known about Elara, which is at a distance of 11,737,000 km from Jupiter's surface and rotates every .5 days with an orbital period of 259.6 days. Comparing these numbers, as the orbital period increases, the rotational period increases. Differences are due to orbital eccentricity for Himalia of .1580 and the orbital eccentricity for Elara is .2072. What is unusual is, despite the different shapes of the bodies, they are in similar orbit conditions but they also seem to have the same angular momentum.

Most of the asteroids have been closely observed seem to also be rotating. Unless the capture mechanism involved a collision, it is hard to know how the angular momentum would change by the capture process. It is possible these two moons were rotating before they were captured and maintained this rotation rate while being captured but this would be fortuitous. The rotation rate should be totally random. After capture, tidal effects, or whatever process that creates synchronization, would slowly reduce the rotation rate transferring the angular momentum to Jupiter. Light can carry angular momentum, so heat from tidal friction could radiate some of the angular momentum into space. In other words, two satellites with almost equal weight but different geometric shapes tend to orbit Jupiter at similar distances and have similar rotation rates implying Jupiter's gravitation is inducing angular momentum. Interestingly, the different shapes would result in different inertial matrices that should influence rotational rates. This is insignificant. This has to be more than an unusual coincidence.

C. *Speed of Jets from a Black Hole*

A black hole [9]-[11] may be a collapsed star where the forces of gravity are so large where even light does not escape. When super novas occur, they are the explosion of a star and may result in either a neutron star or a black hole. This means for a black hole, everything that moves at or less than the speed of light will remain within the black hole. This also includes magnetic and electric fields. If a jet leaves the black hole, it must either move at greater than the speed of light or by some other unknown mechanism. The conventional wisdom implies jets are not created by the black hole at all but by matter within an accretion disk and the jets may leave at sub-light speed. The flaw here is the black hole's gravitational pull should be so immense, then the jet matter should fall back toward the black hole.

If greater than light speed, then the jets are clearly evidence of naturally occurring hyper-light phenomenon. Thus it would be beneficial to find validation for a black hole that is not rotating or with no accretion disk but possesses a sole jet. Furthermore, the next question is to understand how much faster than the speed of light exists for such a jet and is there some limit by Mother Nature for the jet velocity?

One may argue the jet consists of debris possessing inertial mass from the surrounding accretion disk. Some black holes may have a jet but no accretion disk. This is a strong possibility; however, if from the black hole itself, the jet consists of a spiral moving outward along its ejection axis as well as rotating about the ejection axis to create angular momentum. Where did the angular momentum come from? The only force that exists is gravity. Is it feasible the linear momentum of the jet is being converted into angular momentum to slow down its speed?

Unfortunately, there are no direct means to measure the rotation rate of a black hole or for that matter, the jet rotation rate. Such rotation can be approximated only indirectly by examining the effects within the surrounding environment. However, it is still quite difficult to accurately perform such an accurate task. This is not the case for a neutron star where the beam of radiation would sweep over the Earth.

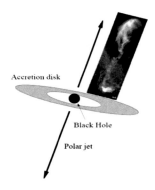

Figure 1. Scheme of a Herbig-Haro object HH47 and the collimated jets of partially ionized gas claimed as belonging to the accretion disk, taken by the Hubble Space Telescope.

Jeong [3] implies the jet from a black hole or a neutron star may be forced outward by a repulsive gravitational source. It is hypothesized the spiral motion within the jet can create a repulsive gravitational source based upon an analysis of a Magnetar by Winterberg [1], [2]. Here we can assume the jet swirls at such a rotation rate if a repulsive gravitational source may result. Moreover, convergence of cloud material could be nothing more than gas dynamic shocks coalescing due to deceleration of the jet matter caused by the original backward gravitational pull of the black hole.

Under the gravitational force magnitude interpretation, an object with negative mass would repel ordinary matter, and could be used to produce an anti-gravity effect. Alternatively, depending on the mechanism assumed to underlie the gravitational force, it may seem reasonable to postulate a material capable of shielding against gravity or otherwise interferes with a gravitational force.

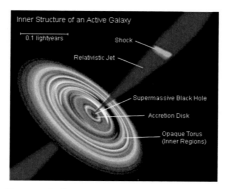

Figure 2. Some details may indicate a gas dynamic shock may appear in the jet as suggested in this artist rendition.

A Magnetar is a neutron star with an extremely strong magnetic field generated by the convection of hot nuclear matter produced as a consequence of nuclear reactions. Winterberg [12] looks at a laboratory analogue of a geodynamo or Magnetar, for a rapidly rotating liquid metal:

$$\nabla \cdot \bar{g} = -4\pi G \rho = 2\omega^2, \quad \text{where} \quad \rho = -\frac{\omega^2}{2\pi G}. \tag{1}$$

According to Winterberg, the source term ρ is negative for a repulsive mass density. If a gyroscope is placed at 45° on a table and let go, the gyroscope obviously falls. However, if the rotor is spinning, it is capable of remaining aligned at this initial angular orientation. As the rotor speed decays, the gyroscope starts to precess rotating in a circumferential direction. When the rotation drops below a certain limit, the gyroscope falls to the tabletop. The conventional wisdom suggests angular momentum couples exist to explain this effect. An alternative solution is the rotation may induce a repulsive gravitational source to levitate the gyroscope according to this equation. Moreover, one could classify this as an 'Inverse Coriolis effect'. Another way of considering this is a gravitational field would repulse negative mass. Such a source can be considered as dark matter [13].

Now the problem is not one where we are suggesting where the entire gravitational contribution due to the neutron star is repulsive. What is suggested is the gravitational contribution due to the mass of the neutron star is partially reduced by the effect identified by Winterberg assuming this is true. The other possibility is if the mass gravitational component is insignificant, then obviously this may provide the means for a Propulsor system. This is why the effect should be thoroughly examined.

D. Pulsars

A pulsar or neutron star is the collapsed core of a massive star, which ends its life in a supernova explosion. When this occurs, there are two possibilities of creating either a black hole or a neutron star. Let us only consider the latter. Weighing more than our Sun, the star collapses because of gravitational forces to become only about 20 kilometers across. As the star collapses, conservation of angular momentum indicates the rotation rate should increase. The collapsed star has an intense magnetic field that may not align with the rotation axis. These incredibly dense objects produce beams of radio waves, as well as X-rays, sweeping a swath across the sky like a lighthouse, often hundreds of times a second. Radio telescopes receive a regular train of these pulses as the beam repeatedly crosses the Earth so the objects are observed as a pulsating radio signal. Pulsars make exceptional clocks, which enable a number of unique astronomical experiments. Some very old pulsars can reach rotational speeds of over 600 rotations per second, rotate so smoothly where they may "keep time" more accurately than the best atomic clocks here on Earth. Thus, the implication is that with the proper software and detectors, pulsar timing would be more accurate than GPS.

One can postulate the formation of a binary pulsar in the strong gravitational attraction would pull another celestial body within the vicinity of the Neutron star. For binaries, there can be a flow of material into the neutron star from a companion star. However, analysis to date tends to suggest that for a companion and the attracting neutron star, the weight is generally comparable. This is unusual because weight differences for these bodies should be random and would impact the types of orbits about each body. If one is to accept Jefimenko's notion where gravity induces angular momentum, then there should be an effect which impacts the rotation rates of both the primary and companion star. Unfortunately, there is no technique available to determine the rotation rate of the companion star in a pulsar binary system. Moreover, regarding Winterberg [12] where rotation creates a repulsive gravitational force, the two bodies may simulate a binary system where each body appears to have comparable weights by balancing the rotation rate of the neutron star. If gravity changes angular momentum, then the companion will affect the rotation rate of the neutron star with a careful balance occurring between rotation rates and the subsequent orbits of both bodies.

If Winterberg is correct, then the inertial mass of the neutron star has to be greater than the companion star to compensate for the loss of gravitation due to spin. In other words the neutron star should weigh more and the value depends upon the rotation speed. Neutron stars at 600 rps would produce more compensation compared to a neutron star that rotates at about 10 rps. The neutron star source term is greater by:

$$\rho_p = \rho_c + \frac{\omega_p^2}{2\pi G}, \quad \text{and} \quad W_p \approx \left[1 + \frac{\omega_p^2}{2\pi G \rho_c}\right] W_c \qquad (2)$$

The subscript 'p' is for the neutron star while the 'c' is the value for the companion star. The neutron star may be located at the epicenter of the elliptical trajectory and the companion star essentially orbits about the neutron star. Such a trajectory is more conventional for the two-body problem of space mechanics as one possibility. A second possibility is both bodies travel in an elliptical orbit based upon their mutual attraction for each other. A third possibility, which is closer to the truth, is that these bodies rotate about different elliptical orbits. Thus, the suggestion is we may have underestimated the weight of celestial bodies and the additional weight may negate the need for finding additional mass thereby competing with such hypotheses as dark matter to explain anomalous phenomenon [14] regarding cause and effects.

Differences in masses may dictate the neutron star spin rate as an artifact of a natural gravitational process where gravity is not only an attractive force but induces angular momentum per Jefimenko's claims. This formula suggests the only case where a Neutron star has the same inertial weight as its companion is when there is *no* Neutron star rotation. No observational data for such a situation exists, which without the neutron star beam sweeping consequences produced by the lighthouse effect would essentially defy detection.

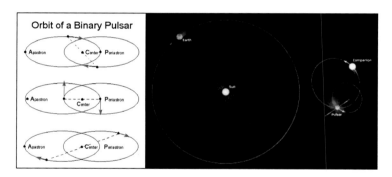

Figure 3. The orbits of the binary pulsar 1913+16 and a comparison of the orbits of the pulsar J1903+0327 with its possible sun-like companion star with the orbit of the Earth around the sun [12]. Thus if weight was the same, one would assume to have similar orbits.

Ideally, consequences of these interactions would produce a circular orbit that contains both bodies. However, this is not the case and these binary pulsars contain either a single elliptical orbit or each body would travel in its own separate elliptical orbit. Moreover, with the passage of time, mass from the stars decreases due to emission of energy in the form of X-ray, electromagnetic, or gravitational wave radiation. Thus the decreasing masses, the orbit shapes, and the rotation rates all have to reach certain equilibrium values to result in an elliptical orbit and with more expenditure of masses from both stars, eventually reach a circular orbit when the masses become comparable.

IV. RESULTS

As the binary pulsar evolves for two bodies, it goes from two erratic elliptical orbits and may eventually collapse into a single orbit. This is what one would expect but it does not happen. With time, considering that the mass decreases in the neutron star, the single elliptical orbit evolves into a single circular orbit. If this is the correct evolution, the propulsion implications are obvious. If true, then clearly rotation or angular momentum has a direct relationship with gravity and Jefimenko was right in his conjecture. Thus, there is a direct connection between angular and linear momentum. Moreover, Winterberg would also be correct in his notion where rotation can induce a repulsive gravitational effect and this also has propulsion implications.

A. Mass Determinations

One of the interesting points about this is the determination of the weight of the neutron star. Some of this data is based upon experimental data and the way to find this compared with the weight of the companion is addressed at a later chapter. However, the analysis, which currently exists has data as follows for several binary pulsars:

Table I. These weights and some of the analysis are provided by [15]-[20].

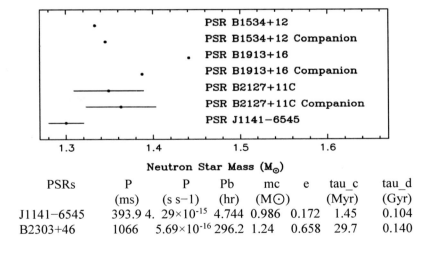

PSRs	P (ms)	P (s s−1)	Pb (hr)	mc (M☉)	e	tau_c (Myr)	tau_d (Gyr)
J1141−6545	393.9	4.29×10^{-15}	4.744	0.986	0.172	1.45	0.104
B2303+46	1066	5.69×10^{-16}	296.2	1.24	0.658	29.7	0.140

B. Trajectory Parameters

Table I has some interesting data. Each mass is in solar mass or the weight of our sun. Earlier it was suggested the Neutron star should weigh more than the companion star. Differences shown here may indicate in these binary systems, the companion star may also be another Neutron star; however, without the lighthouse effect, detection is denied with no other conclusions for the companion being a neutron star.

Other information on mass is from:

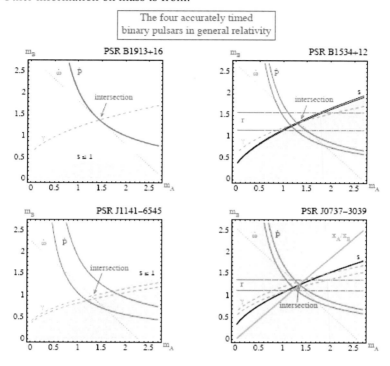

Figure 7. Mass comparisons of several binary pulsars where the Neutron star and companion star are approximately close to the same weight.

Data was taken to evaluate several binary pulsars to establish some insights regarding the size of the primary body, a Neutron star, and its companion. This data shown has some uncertainty by virtue of the observational or radiometric data considering the large distances involved. Moreover, the data set itself is incomplete because of the numerous sources of which most is from Wikipedia includes claims where data flowed from peer-reviewed technical papers. Where applicable, uncertainties in the data are included. In some cases, the authors imply the binary pulsars are moving in nearly perfect circular orbits; it shall be assumed if not mentioned, the eccentricity will be close to zero. However, as shown for Pulsar 13 (PSR J1903+0327), these two bodies may traverse circular orbits but both bodies are not in the same orbit. Note further the masses of both bodies are not equal.

Table I Pulsar Information

Pulsar	Eccentricity ϵ	P(ms)	$M_{primary}$	M_{comp}	Rotation Rate	Orbital Period
1. J1141-6545	0.17	393.93	1.30±.02	.986±.02	2.538Hz	4.77hr
2. B2303+46	0.66	1066	1.40(?)	1.24	0.93Hz	296.2hr
3. PSR J0437-4715	0.00	5.75	1.8	.25	173.7Hz	5.75d
4. PSR J1518+4904	0.25	40.9	1.75	0.9-2.7	24.44Hz	8.6d
5. PSR B1534+12	0.2736	37.0	1.339	1.339	26.38Hz	.4207d
6. PSR J1811-1736	0.828	104.2	2.6±.9	0.7	9.59Hz	18.8d
7. PSR B1820-11	0.79	279.8	----	----	3.58Hz	357.8
8. PSR B1913+16	0.61713	37.904	1.4414	1.3867	26.3Hz	.3229
9. PSR B2127+11C	0.68	30.5	1.34±.2	1.365±.2	32.7Hz	.3
10. PSR J0737-3039						
Pulsar A	-----	22.7	1.337		44.04Hz	50sec
Pulsar B	-----	2.77sec	1.250		0.36Hz	.3667s
11. PSR B1534-12	0.2736	-----	2.67838	1.339	-----	.4207d
12. PSR J1748-2446ad	0.00	1.39	~2.0	.14	716Hz	26Hr
13. PSR J1903+0327	0.437	2.15	1.74	1.0	465Hz	95.17d
14. PSR B1937+21	-----	1.5577	-----	-----	641.9Hz	-----

This is part of the problem that makes the celestial mechanics of binary pulsars so interesting. Things appear to change in unexpected fashions. With this anomalous behavior, there should be some intellectual growth to understand this phenomenon. The issue about mass differences between the neutron star and the companion within the binary system is a major concern.

Pulsar 5 (PSR B1534+12) and Pulsar 9 (PSR B2127+11C) are where the Neutron star and companion body are very close to the same weight. Typical masses are from [19]. Note the error bar from some of these values. However, these are in elliptical orbits with eccentricity of .2736 and .68 respectively. The latter is considerable. However, note also the weights for each binary star are comparable. The only difference for these binaries is the rotation rates of the Neutron stars. It is assumed the orbits of both bodies in each of the binaries with similar orbits. The point is if the weights of the Neutron star and the companion are close to each other, then they should be in circular orbits. If a new capture, the orbit should decay to one with slight eccentricity. This is not the situation by any stretch of the imagination. One could assume the rotation rate of the Neutron star creates an imbalance for such orbital kinematics.

There is a contrasting situation as well. This is with Pulsar 3 (PSR J0437-4715) [19], [20] and Pulsar 12 (PSR J1748-2466ad) where the eccentricity is zero and it is assumed the neutron star and companion are in the similar circular orbits unlike the situation shown for Pulsar 13 (J1903+0327). Optical observations from several references indicate the binary companion of Pulsar 3 (PSR J0437-4715) is most likely a low-mass helium white dwarf. The pulsar is about 1.8 solar mass and the companion is only about 0.25 solar masses. The pair revolve around each other every 5.75 days in nearly perfect circular orbits. This would imply residing in a circular orbit, both the weight of the neutron star and the companion star should be equal. However, as shown, the mass of the neutron star is almost an order of magnitude greater than that of the companion star. Moreover, the rotation rates are quite close as well. This implies the rotation rate may potentially decrease the inertial mass quite a considerable amount to match the inertial mass of the companion star. If true, this suggests rotation effects may create a situation, which results in a component having a repulsive gravitational source.

If these masses for the binaries are correct, this is a significant finding for creating a future space propulsion capability. In other words, our spacecraft may have to replicate and duplicate the characteristics of a neutron star in a binary pulsar if we are to become a space-faring civilization and leave the blue marble.

V. CONCLUSIONS

Intuitively, we may have accidentally discovered the Holy Grail of space travel. Converting angular momentum to create linear momentum can occur in space when if it goes through an intermediate step, say inducing a gravitational repulsive source. This may be demonstrated by rotating swirls of inertial jets to produce a repulsive source strong enough to leave the gravitational field of a black hole. This will create a forward acceleration or enough linear momentum to leave the black hole. Obviously, there is a strong want to validate this conjecture with experimental evidence. Moreover, binary pulsars may appear to enjoy a careful balancing act between the primary and companion star whose masses can be significantly different. These stars use gravitational angular momentum effects to determine the neutron star's rotation rate. Based upon these characteristics, subsequent orbits will initially be separate elliptical orbits that slowly evolve into similar elliptical orbit. Thus when the 'virtual' masses of both stars are nearly equal, either by consuming neutron star mass to produce energy over time or change the star's rotation rate, the orbit of the stars should circularize into circular orbits. Now the hard part, *'How do we convert this knowledge, if true, by creating hardware for a future space drive to take mankind to new star systems?'*

Acknowledgments

The author wishes to acknowledge information received from John Cole, formerly of NASA Marshall Spaceflight Center, regarding the Moons of Jupiter. The author wishes to also thank Mr. Dana Sweet for helpful criticisms. The author appreciates comments made by the reviewer.

References

[1] Murad, P. A., An Anzatz about Gravity, Cosmology, and the Pioneer Anomaly, in Proceedings of *Space, Propulsion and Energy Sciences International Forum*, edited by G.A. Robertson, AIP Conference Proceedings **1208**, Melville, New York, 2010.

[2] Murad, P. A., The Challenges of Developing the Technology for A Realistic Starship Propulsor, presented at the 48th AIAA Aerospace Sciences Meeting, AIAA Paper 2010-1609, **2010**.

[3] Jeong, E., An Isolated Gravitational Dipole Moment Placed at the Center of the Two Mass Pole Model Universe, arxiv.org, 1997.

[4] Murad, P. A., Warp-Drives, The Dreams and Realities, Part I, A Problem Statement and Insights, in the proceedings of the *Space Technology and Applications International Forum (STAIF-05)*, edited by M. S. El-Genk, AIP Conference Proceedings **746**, Melville, New York, 2005, pp. 1256-1263.

[5] Murad, P. A., Warp-Drives, The Dreams and Realities, Part II, Potential Solutions, in the proceedings of the *Space Technology and Applications International Forum (STAIF-05)*, edited by M. S. El-Genk, AIP Conference Proceedings **746**, Melville, New York, 2005 p. 1411-1418.

[6] Jefimenko, O. D., *Gravitation and Cogravitation*, Electret Scientific Company Star City, 2006.

[7] Murad, P. A., An Alternative Explanation of the Binary Pulsar PSR 1913+16, in Proceedings of *Space, Propulsion and Energy Sciences International Forum*, edited by G.A. Robertson, AIP Conference Proceedings **1103**, Melville, New York, 2009.

[8] Imagine.gsfc.nasa.gov/docs/science/know_12 /pulsars.html.

[9] Novikov, I., *Black Holes and the Universe*, Cambridge University Press, 1995.

[10] Visser, M., *Lorentzian Wormholes from Einstein to Hawking*, AIP Press, 1996.

[11] Thorne, K. S., *Black Holes & Time Warps-Einstein's Outrageous Legacy*, W.W. Norton & Company, New York, London, 1995.

[12] Winterberg, F.: *Thermonuclear Dynamo inside an Alfven Black Hole*, University of Nevada, Reno, 2006.

[13] Brandenburg, J. E. and Kline, J. F., Application of the GEM theory of Gravity-Electro-Magnetism Unification to the Problem of Controlled

Gravity, Theory and Experiment, presented at the 34th Joint Propulsion Conference & Exhibit, AIAA 98-3137, 1998.
[14] Murad, P. A., Gravity Laws and Gravitational Wave Phenomenon, Is There a Need for Dark Mass or Dark Energy? Presented at the AIAA/ASME/SAE/ASEE Joint Propulsion Conference, AIAA 2008-5123, 2008.
[15] Courtesy NASA/JPL/University of Arizona.
[16] Saxton, Bill, NRAO/AUI/NSF
[17] Dunham, W., Astronomers Baffled by Weird, Fast-Spinning Pulsar, *Reuters*, Washington, DC, 2008.
[18] M. Bailes , S. M. Ord , H. S. Knight , and A. W. Hotan , *Centre for Astrophysics and Supercomputing, Swinburne University of Technology, P.O. Box 218, Hawthorn, VIC 3122, Australia; mbailes@swin.edu.au. Received 2003 July 3; accepted 2003 August 7; published 2003 August 25.*
[19] Bell, J. F., Bailes, M., Bessell, M. S., Optical detection of the companion of the millisecond Pulsar J0437 – 471, *Nature* 1993 **364**(6438): 603-605.
[20] Verbiest, J. P. W., Bailes, M., van Straten, W., Hobbs, G. B., Edwards, R. T., Manchester, R. N., Bhat, N. D. R., Sarkissian, J. M., Jacoby, B. A., Kulkarni, S. R., Precision Timing of PSR J0437-4715, An Accurate Pulsar Distance, a High Pulsar Mass, and a Limit on the Variation of Newton's Gravitational Constant, *The Astrophysical Journal* (2008) **679**(1): 675-680.

Nomenclature

c Speed of light (3×10^8 meters/sec.)
g Gravity (9.8 m/sec^2)
G Gravitational constant
m Mass (kilograms)
p Pressure

Greek Symbols

ε permitivity (farads/meter)
μ Permeability (henrys/meter)
σ Conductivity (mhos/meter)
Λ Cosmological constant
γ Relativity factor
ρ Volume (kilograms/m^3)

Chapter III

An Assessment Concerning Neutron Stars and Space Propulsion Implications

Abstract: There are many unknowns in relation to stellar evolution of neutron stars. Neutron stars might possess multipolar architecture in lieu of a single dipole claimed by the conventional wisdom. The multipole issue cannot be resolved using a single point observer reference point such as the Earth, but would require a non-terrestrial observer location with a significant offset. Moreover, the question is how a neutron star's magnetic field may be created considering differences between the neutron core and a gas surface layer of protons and electrons. These differences between the layers constitute charges with moving currents that result in a magnetic field supported by a fast moving rotating core. If electrons in Cooper pairs exist in a neutron star, then the amount of magnetism may increase by acting analogous with superconductivity. By symmetry, proton pairs should also be real to produce similar charge redistributions. With these thoughts, a laboratory replications of a neutron star is proposed to create a large magnet that may be suitable for a space mission.

PACS: 04.50.Kd, 04.80.Cc, 06.30.Dr, 06.30. Gv, 97.10.-q, 97.10.Gz, 97.10.Xq, 97.60.Gb, 97.60.Jd, 97.60.Lf, 97.80.-d.
Keywords: Binary Pulsar, Neutron Star, Asteroid, Gravity, Jefimenko, Rotation, Angular Momentum, Trajectories.

I. INTRODUCTION

There is a necessity to discover numerous secrets about neutron stars and binary pulsars to gain possible insights about new types for creating energy or creating a large magnet. Pulsars are mentioned because they "pulse" with emissions at a steady observable rate. For example, PSR 1919+21 pulses every

1.337 seconds. These pulsars rotate incredibly fast, throwing off beams of energy that sweep across the Earth using radio waves or X-rays from their magnetic poles. Knowledge about neutron stars within pulsars would allow us to understand the far-field environment of space and possibly enable the creation of new propulsion system concepts. Let us first examine some background information to demonstrate our basic understanding of a neutron star before discussing possible propulsion or energy implications.

Most of the magnetic fields of cosmic objects are generated and maintained by the dynamo action using electrically conducting fluids in motion [1]. Essential features of the dynamo theory of cosmic objects are developed, first on the kinematic level and later taking into account the full interaction between magnetic field and motion. Particular attention is paid on electrodynamics as well as magnetofluid dynamics with application to dynamo models for surface objects on a planet showing irregular or turbulent motion and magnetic fields. A few explanations are given by acting on such a single dynamo with the Earth and the planets, in the Sun and stellar objects and in galaxies.

Magnetic fields of cosmic objects show a great variety [1] not only with respect to their magnitudes and spatial extents but also to their geometrical structures and time behavior. As far as our Sun is concerned, not only sunspots, but also all phenomena of solar activity such as flares, protuberances, and coronal mass ejections are connected with magnetic fields. These magnetic properties are measured with the help of the Zeeman Effect [2], [3] that splits a spectral line into several components in the presence of a static magnetic field.

From the study of sunspots and related phenomena of the solar activity cycle, we may conclude the Sun possesses in general a large-scale magnetic field [4] that consists mainly of two field belts beneath the visible surface with flux densities exceeding at least 10^{-1} Tesla [1]. This belt is in the northern hemisphere and the other exists oppositely oriented in the southern hemisphere. In addition, there is a much weaker poloidal field with only a few 10^{-4} T intersecting the visible surface. Clearly, the near surface magnetic field is not easily established but rather complex; however, a stronger dipole relationship endures for the far-term view in our solar system.

After the discovery of the pulsar phenomenon [1] in the late sixties, the only acceptable explanation for a pulsar could be given by assuming a rapidly rotating neutron star with a very strong non-symmetric magnetic field, which is oriented that is misaligned about the rotation axis. Thus this field is oriented as an oblique rotator. From the observational data flux densities of the order of 10^8 T, these fields for neutron stars were derived. In a few cases the existence of such strong fields has been confirmed in an independent way by the interpretation of X-ray spectral features due to electron cyclotron resonance scattering. Recently the observation of anomalous X-ray pulsars suggests there are even enough neutron stars with flux densities as large as 1,012 T.

II. DISCUSSION

The objective here is to understand how a neutron star may exist based upon our own sun. There are many theories about how a neutron star collapses. Let us assume we look at our star with a given initial rotational axis. There are several magnetic fields on the sun as previously mentioned in Figure 1. The poles may not be aligned in a given dipole axis. Here, it is not if a clearly distinguished north or south magnetic pole can be normally expected. The classical theory of electromagnetism offers two causes for magnetic fields: permanent magnetization of condensed matter and/or moving electric charges can become currents. Conditions allowing permanent magnetization can be excluded for almost all cosmic objects by several reasons. Electric currents in conducting matter are subject to Ohmic dissipation, which converts the energy stored in a magnetic field into heat. If there is no electromotive force that maintains the currents and so to compensate this energy loss, the currents and the magnetic field are bound to decay.

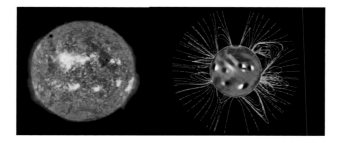

Figure 1. The first picture is the Venus transit across the sun on 5 June 2012. Note the solar flares are anchored to the edges on the surface acting as independent magnetic field poles are not in a singular pole as observed on Earth.

In contrast to the Earth, where the kernel [5] lies inside the iron core, the material at the center of the sun resides in the state of an ideal gas. Our sun consists of several different or multiple poles throughout the sun's surface. This implies each field is bound upon itself. These poles may be considered as 'minor' poles. Moreover, our star has an unusual magnetic field, which is far stronger and different than expected. The impact upon the far field of the sun's magnetic field results in a growing increasing spiral.

This data is shown on Figure 2 based upon instruments over decades from Pioneer 10 and 11. This magnetic field from considerable distance appears where the sun acts as a singular pole in lieu of multiple poles observed on the sun surface. Since these satellites discovered a gravitational anomaly as well as demonstrated with three other long-range satellites, the data may support the contention that gravity can be altered. These Pioneers may occur when a satellite passes through the spiral's magnetic trenches to influence long-range gravitation. Thus, there will always be some unexpected events with gravity as we get exposed to further distances from the Earth.

III. AN ASSESSMENT

Let us address the evolution of a neutron star, which is initially born with a star similar but larger than our sun. The issue is to understand this phenomena to use this knowledge for creating a better model and potentially a space propulsion scheme.

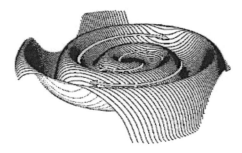

Figure 2. The sun magnetic field based upon a spiral is based upon measurements made by the Pioneer 10 and 11 where each moved in opposite directions within the solar system. (*Image is from J. Jokipii, University of Arizona.*)

A. Pulsar Models

There are many uncertainties in stellar evolution [3]. Let us examine a basic star similar to our sun, which will undergo a process that leads toward creating a neutron star. The conventional wisdom suggests a neutron star occurs after many stages of nuclear burning of the star core. The core becomes unstable and implodes while the outer layers of the star explodes as a supernova. There is a massive output of energy. Neutrons are created from decomposition of the core moving through the outer layer and nucleosynthesis for heavy elements. Outer layers of the gas are thrown off and become some of the supernova remnants. The subsequent core explodes and may leave a remnant such as a neutron star that consists mostly of neutrons. The neutron star rotates rapidly observable as a radio pulsar. It is possible if the remnant may either create a neutron star or alternatively create a black hole.

The structures of the neutron stars more closely resemble a planet than a star [6] but its atmosphere is on a smaller denser scale. The atmosphere, if any, is unlikely to be more than a few centimeters thick [7]. The outer layer of the star is considered as a crust that is solid and crystalline and probably several centimeters thick. This crust consists of a neutron cloud or superfluid containing a few protons and electrons to stabilize the environment.

In some cases neutron stars [8] are expected to slow down such as PSR 1913-16 where the difference in mass and energy is believed to be converted into gravitational waves. In some cases, the star can actually spin up with X-ray stars, which are unlikely radio pulsars without power by their stored rotational energy. Neutron stars are alone in interstellar space and capture material from its surroundings. Moreover, since supernovas are rare events, we must conclude most stars will die peacefully for the transformation into a neutron star.

Figure 3. The Crab Nebula represents the first neutron star is expended with a large amount of energy.

B. A Contrasting View

Let us look at this differently. Neutron stars are believed to be a product of a massive star after a supernova. The first thought that would arise is if there are enough supernovae who exist to satisfy the number of existing pulsars. Basically at least 200 pulsars survive and astronomers with the number of supernovae that have discovered more than the lower number where, for example, four of which subsist alone in the Milky Way. However, the star is not big enough to create a black hole but is large enough to create a tiny, dense and hot mass of neutrons. Intense gravity forces the matter into a tiny volume but quantum effects repel the neutrons. At the center of the Crab Nebula lies the Crab Pulsar, a neutron star, 28–30 km across, which emits pulses of radiation from gamma rays to radio waves with a spin rate of 30.2 times per second. The nebula in Figure 3 was the first astronomical object identified with a historical supernova explosion. The question is the geometry of the neutron star. In other words, how could a neutron star result in a spherical shape during the supernova's implosion?

C. The Birth of a Neutron Star

If an explosion occurs on the star's surface to create an implosion, this would compress the central core. However, the explosive process would require the entire region of the star to be evenly distributed covered for compression to produce a spherical-like object. This probability for a star for this implosion is most likely an ovoidal-like geometry or possibly contains topological ridges. The probability of a perfect sphere should be viewed as having a very low or none probability and the differences in geometry may produce unexpected uneven topological effects. Likewise, the notion that the probability of the surface is smooth may also be impractical. This could have an impact on the onset of turbulence when the gas boundary layer of the atmosphere interacts with the rotating core. Furthermore, a large portion of the star would separate outwardly due to the compression process or explosion. The resultant rotational rate would count upon available mass for angular conservation due to a reduction in the moment of inertia of the initial star. This obviously includes a minus portion of mass moving radially outward during the supernova explosion.

There is an issue of how a star becomes unstable, most likely due to magnetic field changes, to produce either a neutron star or a black hole. What are the criteria and the processes? If the star becomes unstable, it most likely occurs on the surface for supporting an implosion. How does the surface allow the instability to propagate over the entire surface before an implosion occurs to produce a near-perfect sphere? If such an instability moves radially from an initial point, inhomogeneous density variations will exist within the initial star's interior. The mechanism for only allowing a surface disturbance is not well-understood or addressed because an implosion must be perfectly timed but the probability should be low for this to occur.

If the instability occurs at the center, all of the star should move outward radially to probably form an accretion disk and the result would most likely be a black hole.

There is another issue. Let us assume the star explodes and the explosion is a spherical shell. The star will have a distribution based upon the initial pressure and gravitational distribution. Obviously the center should have the largest density because of gravity. If there is a concentric layer of different particles or elements in the initial star interior to become unstable for an explosion, it could be unevenly distributed to produce a constant density neutron sphere. However, if the layer has a specific thickness, it could form the explosion with an offset to compress the star with considerable differences in density distributions. Here, the possibility of generating a perfect spherical core geometry should be extremely low based upon these density variants. Moreover, these asymmetric density structures may create unusual dynamic responses similar as seen in

Figure 3. In other words, if the expanded cloud was spherical, then the neutron star evolution would be perfectly spherical because of the 'perfect' implosion. Since it is not, then the conclusion is the neutron star may not be perfectly spherical. In fact the shape of the cloud should give some insights about the shape of the neutron star. For example, this would be an ovoid with one axis almost twice as long as the other axis. Such a difference will allow a wobble where the magnetic field beam may not coincide with the rotational axis.

The other part of the conventional lore is the atmosphere is very low. The neutron core does not create an electric charge and will not induce magnetism unless there is free motion of electrons or protons within a high-density neutron matrix. The separate layers of electron and proton gases [7] are probably the prime ingredients to induce a large magnitude magnetic field. This will induce an electromagnetic effect from eddies and vortices that will spin in closed loops attached to the surface depending upon surface sharp edges or ovaloidal-like geometries. Each of these layers will create charges and the diffusion between the layer interfaces will induce moving currents to result in magnetism. This may involve Taylor instabilities of the boundary layer in the proton and electron gas layers to support magnetism.

Another approach to create a neutron star would be with a more gradual change or a slow crunch. This would be valid if no supernova remnant debris exists. As the growth of the gravitational attraction starts the collapse, the central core in the star would pull under hydrostatic and gravitation pressure to push together electrons and protons available in an atom to form neutrons. If gradual, you would not observe any remnants of the supernova. Moreover, this layer of the neutron core is due to gravitational fields to grow as the neutron core density layer increases in an outward radial direction within the remnants of the remaining portion of the reducing star. The size of the star collapses in volume and compresses the core into a smaller overall volume. It is difficult to determine if the changes in the star during this process would increase or decrease the initial strength of the existing gravitational and magnetic fields by conservation; however, if you look at the gravitational source term, the density should increase in strength with the formation of densely packed neutrons.

The density concentration within the core throughout the neutron star could have two possibilities; the concentration could be either a constant distribution layer of neutrons within an evenly distributed crystallize structure or they could obey as a matrix as an inverse radius relationship due to pressure differences throughout the interior core toward the surface. This distribution needs to adjust the effect of centrifugal force due to the high rotation rate. Such an effect will counter some of these gradients.

From the short-range nuclear force between neutrons, this could be suspect of a near-constant density if thermal effects and the Pauli Principle are ignored. Neutrons would try to minimize the Pauli effects by exhibiting crystalline structure, at least in local volumes on the surface region. Thus, the neutron must align itself due to a pressure-like relationship.

Assuming the entire star consists of neutrons, how is the magnetic field created along a detected singular axis? Recall our sun has many poles and how is this reduced to a singular pole at significant distances? In this model, a celestial body consisting solely of neutrons should not allow a magnetic field. The rotation of the neutron star would conserve the initial star's angular momentum. However, differences considering making measurements at long distances from the Earth are always problematic and tainted by uncertainty. This suggests observations could be distorted due to gravitational lensing either by other stars or galaxies that intercept between receiving the pulsar seen on the Earth. Finally, the effects observed by a neutron star's magnetic field would be similar as in Figure 2 for the sun except the spirals would be closely spaced due to the high rotation rates and extend further due to the higher magnetic field.

D. Pulsar Lighthouse Effects

In time, either violently or more likely during gradual evolution, the star crunches down from an average size of about 1.25 times the size of our sun to decrease down to a diameter of about a 10 to 20 km. To do this, conservation of angular momentum determines some of the neutron star's initial rotation rate, say from initially at ten times a month, to increase from about 10 to 600 revolutions per second. When one observes the neutron star from the Earth, a beacon may be observed as a lighthouse effect that sweeps across the Earth as shown in Figure 4.

Figure 4. The Pulsar Lighthouse Effect

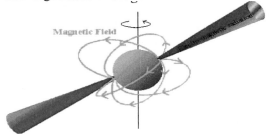

Obviously there can be possible situations where a neutron star exists but the beacon does not sweep or is not observable as a beacon on the Earth. This could be a situation where we see a supernova but there is no detection of a neutron star because of no lighthouse sweep. Thus a neutron star may rotate with an axis along the 'major' dipole magnetic field misaligned from the rotational axis. Electromagnetic energy is aligned with the beacon due to the magnetic field along the direction of the poles in the neutron star. Briefly, this assumption found about pulsars may be profoundly incorrect; however, this will be found only if we travel freely through the cosmos. Let us address some of these situations.

There are many different possibilities in the collapse process. If the rotational axis and magnetic pole of a neutron star was collinearly oriented, the lighthouse effect would not occur if seen by the Earth but would appear as a continuous beacon; however, a continuous radiation source to date has not been observed as a pulsar unless it was either classified as a Quasar or some other type of unusual celestial stars. The other part of the problem is if more than one beam may persist. This could be determined for multi-beams only if we observe the pulsar at some far distance from the Earth.

Basically this means there may be considerably more discoveries over the existence of the number of pulsars. We may be seeing only 10% of the total that exists.

E. *A Mechanism to create Magnetism*

How does the semblance of neutrons induce a magnetic field if they consist entirely of neutrons? If you look at the Earth, the moving molten core generates a magnetic field around the Earth. The core rotates at a different rate from the Earth's crust and there occurs moving tectonic surface plates where this difference creates currents resulting in magnetism. There is no moving or molten core on the Moon and it only has minute amounts of magnetic ore on the lunar surface. The same is true about the Martian surface where magnetism is based upon only specific regions of magnetic ore. If the neutron star is so densely packed, there is probably no internal motion of the neutron star core. Moreover, if there is nothing more than neutrons, there is no mechanism for creating a magnetic or electric field. Thus, what is the mechanism for creating an electric or magnetic field for a neutron star?

Generating the magnetic field is not trivial. The neutron star could consist of a layer of proton gas covered with an electron gas created by degenerate neutrons. If this were true, then the rotational motion of the collapsed star would have a magnetic field due to the differences between the neutron core, the trapped protons, and a layer of electrons. The proton gas consists of higher density than electrons and would fall underneath the electron layer due to gravitation and buoyancy. However, the high rotation rate could create centrifugal motion forcing convection to transport heavier proton gas further away from the electron gas. This is highly unlikely considering the large gravitational field. Some interactive motion can occur in the interface between both of these layers to generate electric currents and between the layers and, with stationary charges in the layers themselves will produce a magnetic field using Maxwell's equations. If anything, the electron gas most likely will leave the neutron star.

Let us consider the proton and electron gas layers. A loop of proton current can encircle a magnetic flux and be trapped by this flux. This need not happen at a magnetic pole. To escape, the proton and electron currents would have to cut the flux and thereby create an electric field opposing the action. A surface vortex loop would tend to drift on a rotating body due to Coriolis. If the opposing electromagnetic forces overwhelm the Coriolis force, the loop will remain in the same physical place on the surface of the star. As long as the overall strong flux field is fixed on the rotating body, the current loops trapped in the field are dragged around with it.

There is another matter of where the energy is coming from the power in the radiating loops. The gas state [4] could transport eddies of protons and waves of electrons on the surface of the star producing strong electromagnetic waves. These could be thought [7] of similar as the "Red Spot" on Jupiter's surface. Storms of surface proton and or electron gas could carry around trapped magnetic lines of force to remain at the red spot and still become similar to a neutron star's "lighthouse" beacon.

This situation is quite complex. The proton and electron gas will form the magnetic dipole of the neutron star and if so, this is not oriented in the rotational axis of the neutron star based upon edges embedded in the exterior surface of the neutron core or ovaloidal geometry. This would create vortices in the plasma and might naturally align with the proton gas along this axis. There is another possibility.

A neutron star may indeed have quadrupoles or multipoles in lieu of a single dipole model similar to the surface of the Sun and possibly other stars. These poles may be created by overlapping current loops. If the beacons may exist and are oriented in such a manner, these beacons may not be swept or observable from the Earth. If this goes with the belief of a proton and electron layer, the layers would obviously not be smooth but wrinkled as well as with the neutron core surface and the separate layer interface region. Here is a mixing area at the interface with convection and diffusion between the different proton and electron plasma. Moreover, electrostatic and electrodynamic attractions would also involve such attractions. The proton gas would reflect turbulence on the geophysical interactions with ridges or small hills created by the core. Finally, as the Earth's magnetic field and polar orientation changes as a function of time, it is also conceivable the magnetic pole, either for a dipole or multipole may also shift in the neutron star with comparison to the rotational axis. Star quakes if possible, may reorient the geography of the star's topology to allow such motion and actions altering the magnetic fields.

F. Pulsar Timing and Other Options

The rotational rate seen by the magnetic pulse signal, many times per second, may be more accurate regarding timing than an atomic clock used in the GPS system [8]. There is an important challenge to use space navigation to go to the far planets or to the stars. GPS satellites about the Earth will only have usefulness basically near the surface of the Earth and possibly up to the satellites' orbits. Anything higher for accuracy may limit GPS algorithms other than, say satellites above GEO unless severe changes are included within the algorithm. With specific binary pulsar characteristics include a continuous timing mechanism with a neutron star, it is obvious pulsar timing offers a means for accurate space navigation for using the various pulsars observable at various orientations in the cosmos. The algorithm could be similar to a GPS approach where four or more satellites in the field of view are used for GPS accuracy as well as an anomalous situation for 4 different specific pulsars. Using these pulsars could serve a similar navigation function with considerable or have a higher accuracy than observed by an atomic clock signal used within a GPS satellite. This could be used similar to a GPS-like algorithm to use a navigation network outside of the Earth for travelling to movement in a far-term solar system.

G. Other Possibilities for Space Navigation

A Quasar [11]-[12] is also a very energetic and distant active galactic nucleus. Quasars are the brightest and most distant objects in the known universe and show a very high redshift, with an effect of the expansion of the universe between the quasar and the Earth. They are extremely luminous, first identified as being high redshift sources of electromagnetic energy. This includes radio waves and visible light that were point-like similar to stars, rather than extended sources similar to galaxies. This was a prime motivation for accepting the theory of an expanding cosmos. Most luminous quasars radiate at a rate to exceed the output of average galaxies, equivalent to two trillion (2×10^{12}) suns. Such radiation is emitted across the spectrum, almost equally, from X-rays to the far-infrared with a peak in the ultraviolet-optical bands, with some quasars also being strong sources of radio emission and gamma-rays. In early optical images, quasars looked like single points of light (i.e., point sources), indistinguishable from stars, except for their peculiar spectra. Quasars could be implemented within this architecture as reference or touch stones to assist the pulsar-timing navigation process. These tend to inhabit the very centers of active young galaxies and can emit up to a thousand times the energy output of the Milky Way. However, these bodies demonstrate a galactic representation whereas this could be a singular signal source. In other words, observations for the perfect alignment as well as finding multiple poles for a rotating neutron star would be extremely difficult to observe at a single point of observations such as the Earth. Some of these insights and characteristics [5] are discussed in the Appendix.

This requirement may satisfy a quark star or a hypothetical type of exotic star composed of quark matter, or strange matter. These are believed to be ultra-dense phases of degenerate matter theorized to form inside massive neutron stars. On this basis, the neutron star density would not be uniform but dictated by the pressure differences between the surface and the core center. It is theorized that the neutron-degenerate matter makes up a neutron star under sufficient pressure due to the star's gravity, the individual neutrons break down into their constituent quarks – up quarks and down quarks. Some quarks may then become strange quarks and form strange matter. The star then becomes similar to a single gigantic hadron (but bound by gravity rather than the strong force). Quark matter/strange matter is one candidate for the theoretical dark matter as a feature of several cosmological theories.

Statistically, the probability of a neutron star being a quark star is low, so in the Milky Way Galaxy, there will only be a small population of quark stars. Quark stars [11], [12] and these strange stars are entirely hypothetical as of 2011, but observations released by the Chandra X-Ray Observatory on April 10, 2002 detected two candidates, designated RX J1856.5-3754 and 3C58, which had previously thought to be neutron stars. These different cosmic events may be used to augment as guide stones within a conglomerate pulsar timing algorithm.

H. Death of a Neutron Star

Earlier interpretation suggests a supernova creates a neutron star. As the sun is consumed where mass is converted into energy, the gravitational pull will decrease and the planets will move further away from the sun. The question is how does a neutron star evaporate or disappear similar to a normal star especially in a binary pulsar? Is there a mechanism that allows the neutrons in a crystalline structure to separate the external edges where neutrons are converted into a proton and electron to induce nuclear fusion? Does a sufficient element in the electron and proton gas that generates the magnetic field(s) exist over the surface of the neutron star to create sufficient nuclear fusion? In PSR 1913+16 [8]-[10] where the change in trajectory performance implies reduced kinematic energy, the suggestion is the mass of the pulsar undergoes converting energy into gravitational waves. What is the physical mechanism, which occurs to create gravitational waves especially within a neutron star?

IV. RESULTS

There is insufficient data to evaluate the anomalies previously mentioned. Obviously more data and analysis is warranted. It is not questionable at all about using the presence of a large magnet in space and its impact to induce motion for propulsion. Such a device may take advantage of the solar system's magnetic field and act to oppose its alignment as a dipole with respect to the magnetic flow as part of the solar wind. If the view of Gertsenshtein [13] and Forward [14] are correct where magnetism and electricity relate to gravitation as well as the Murad-Brandenburg Equation [15] concerning a Poynting field conservation could induce torsion or possibly a gravitational field, may allow developing a propulsion system to operate. The objective is to use electro-magnetism to convert gravity and place such a large magnet in a far-term trajectory moving toward celestial bodies toward the outer reaches of the solar system or a trajectory moving toward a star.

A. *A Potential Magnetic Model for a Neutron Star.*

With all of these differences, how would a typical model represent a neutron star in a laboratory environment? The question about a perfectly spherical core is crucial. In nuclear explosions, the implosion process requires charges to be uniformly displaced about the fuel. Each charge is equal in strength and these implosions are detonated in time to simultaneously produce the nuclear reaction. If any of these charges fail, chances are high that a jet will squirt outside of the shell and this would result in a reduced energy detonation. For a supernova, the shells are not evenly distributed and sometimes jets result. The implication is the explosion is not evenly distributed and the possibility of resulting with a perfectly spherical neutron star should be considered as extremely low. The shape of a neutron star should reflect the extremities of the supernova cloud.

If a rotating asymmetric body spins in an axis-symmetric fashion, a vortex will be created that would most likely be oriented along the spinning axis. This vortex would consist of electron and proton gases to generate a magnetic field. Since we do not see a beacon where the magnetic field is oriented with the rotational axis, it is safe to assume the neutron core is not perfectly spherical.

Assuming the neutron core creates edges of a neutron crystal or matrix, each of these edges can create eddies or small vortices due to the neutron star's weather based upon the rotational rate. The atmosphere is not a few protons or electrons thick but has to be a considerable amount to create large magnetic fields. Electron Cooper pairs will generate intense magnetism analogous to superconductivity. It is amazing electrons and protons have the same magnitude in terms of charge strength while they have a difference in mass of 1 to 1836.2. Apparently the electrons because of their lower mass have a higher level of efficiency so to speak. If electron pairs exist, it is reasonable by symmetry that proton pairs may also exist in the proton plasma. The layers of the proton and electron gases generate separate charges while their currents, based upon vortices and turbulent transport mixing, will induce large magnetic fields. These electron and proton filaments will produce Taylor instabilities. Some efforts applied for neutron stars, however, are assumed the core is perfectly spherical. This requires some additional thinking to discover the realities of these stars.

These views are different from the conventional lore and indicate each pulsar represents a unique capability based upon the protuberances on the surface topology, which are yet to be established. Although they can be defined generally, the topology covered by the rotation rate as well as presence of a companion or other star(s) further attest to the uniqueness of these cosmic events.

> *B. A Laboratory Representation of a Neutron Star and a Magnet to Create a Space Propulsion Scheme*

The proposed laboratory system to replicate a neutron star consists of a strong magnet that is rather simple. It will consist of a large air-bearing device using a spherical ball manufactured with Neodymium and nickel married in a steel alloy. There will be surface roughness and small fins acting similar to small turbine blades forcing the ball to rotate within a structural casing. Height of the fins will be large enough to operate the ball like an air-bearing. The clearance between the ball and an insulated casing will be large enough with spacing clearance to withstand a very large voltage. The voltages would be set at 511 KV using microwaves or through other means, which are the voltage needed to strip electrons from atoms as well as anomalous behavior as observed by Maker [16], [17].

The gas-levitating ball bearing will operate a gas with low-pressure hydrogen. This will dissociate the Hydrogen into electron and proton gases as separate entities. Jets will induce rotation on the ball concentric with respect to the casing to spin at some speed of say, 600 revolutions per second or considerably higher. The system gas spins the ball and levitates it centered by the casing. When conditions permit, an electrostatic voltage is provided and the magnet should operate as expected. Both the core and casing should act with an alternating electric charge. The moving core with the gas mixture of electrons and protons should create a magnetic field. Moreover, a Faraday cage would be required for biological safety due to the extensive electrical currents and some of the magnetic effects. This may be suitable for future propulsion devices.

V. CONCLUSIONS

In defining how a neutron star may exist, there are several questions concerning the existing knowledge. The logic raises a model where a neutron core rotates in a layer that contains electron and proton gases forming as separate entities. These gases would be the primary ingredients for creating a large magnetic field. Other information about a neutron star suggests using an advanced navigation system for travel past the trans-lunar region based upon measurable performance of several pulsars. Moreover, it is feasible despite these uncertainties, to create a laboratory replicate of a neutron star to use this knowledge to create a large magnet possibly for space propulsion.

Acknowledgment

The author thanks comments made by Dr. Jack Nachamkin, which motivated this effort with conversations and some very creative ideas regarding using an electron gas over the surface of a neutron star.

References

[1] Karl-Heinz Rääadler, The Generation of Cosmic Magnetic Fields.
[2] Wade, Gregg A. (July 8–13, 2004). "Stellar Magnetic Fields: The view from the ground and from space". *The A-star Puzzle: Proceedings IAU Symposium No. 224.* Cambridge, England: Cambridge University Press. pp. 235-243.
doi:10.1017/S1743921304004612.
[3] Basri, Gibor "Big Fields on Small Stars". *Science* **311** (5761): (2006). Pp. 618–619. doi:10.1126/science. 1122815. PMID 16456068.
[4] http://zebu.uoregon.edu/~imamura/122/lecture-6/solar_activity_cycle.html.
[5] E. M. Drobyshevski, "The DAEMON Kernel of the Sun," *A. F. Ioffe Physico-Technical Institute, Russian Academy of Sciences, 194021,* (E-mail: emdrob@pop.ioffe.rssi.ru), *St. Petersburg, Russia.*
[6] L. Motz, The Cambridge Encyclopedia of Astronomy, Crown Publishers, Inc., New York, 1977.
[7] Conversations with Jack Nachamkin (2012).
[8] Hulse, R. A., and Taylor, J. H., "Discovery of a pulsar in a binary system," *Astrophys. J.*, **195**, (1975a), pp. L51–L53.
[9] Hulse, R. A., and Taylor, J. H., "A deep sample of new pulsars and their spatial extent in the galaxy," *Astrophys. J.*, **201**, (1975b), pp. L55–L59.
[10] Hulse, R. A., "The discovery of the binary pulsar," *Rev. Mod. Phy.*, **66**, (1994), pp. 699-710.
[11] 'Quasar' From Wikipedia, the free encyclopedia
https://en.wikipedia.org/wiki/Quasar.
[12] 'Quark star' From Wikipedia, the free encyclopedia http://en.wikipedia.org/wiki/Quark_star.
[13] Gertsenshtein, M. E., "Wave Resonance of Light and Gravitational Waves," Soviet Physics JETP, vol.14, no. 1, pp 84-85 (1962).
[14] Forward, R. L., "Guidelines to Antigravity," *American Journal of Physics*, vol. 31, pp166-170, 1963.
[15] P. A. Murad and J. E. Brandenburg, "A Poynting Vector/Field Conservation Equation and Gravity- The Murad-Brandenburg Equation," presented in STAIF II, Feb 2012, Albuquerque, New Mexico.
[16] Maker, D., *"Propulsion Implications of a New Source for the Einstein Equations"*, in proceedings of Space Technology and International Forum (STAIF 2001*)*, edited by M. El-Genk, AIP Proceeding Melville, NY, 2001, pp.618-629.

[17] Maker, D., *"Very Large Propulsive Effects Predicted for a 512 kV Rotator"*, IAC-04-S.P.10, 55th International Astronautical Congress of the International Astronautical Federation, the International Academy of Astronautics, and the International Institute of Space Law, Vancouver, Canada, Oct. 4-8, 2004.

[18] Paul LaViolette: *The Talk of the Galaxy* (2000).

Appendix A. Pulsar Timing Details

The signal observed by the pulses has specific characteristics. As mentioned, these characteristics are of such higher accuracy than obtained with an atomic clock. In *The Talk of the Galaxy* (2000), astrophysicist Paul LaViolette [18] revives Sagan's speculation. Here is a brief listing of some characteristics found in the current literature and discussed by LaViolette:

- *Time-Averaged Regularity* - Time-averaged pulse contours do not change over days, months, or years. Timing of averaged profiles is similarly precise.
- *Single-pulse Variability* - Timing and shape of individual pulses can vary considerably.
- *Pulse Drifting (certain pulsars)* - Individual pulses occur successively earlier and earlier within the averaged profile ("drifting pulsars"). For certain drifting pulsars, drift rate abruptly shifts in value. Drift may be random with occasional recurring patterns.
- *Polarization Changes* - Polarization parameters vary within individual pulses, but time-averaged profile of polarization is constant.
- *Micropulses* - About half of observed pulsars exhibit micropulses within individual pulses. Micropulses typically last a few hundred microseconds. They may also have oscillatory periods.
- *Pulse Modulation* - Signal strength may wax and wane over a series of pulses. Or this may be seen only when sampling every other pulse or maybe only at particular times in the profile.
- *Pulse Nulling* - Pulse transmissions may be interrupted for seconds or hours. When resumed, varying parameters continue from where their cycle left off.
- *Mode Switching* - More than one stable pulsation mode, with sudden switching between them.
- *Pulse Grammar* - "Grammatical" switching rules.
- *Glitching* - Pulse periods grow at a uniform rate (as though spinning pulsar is slowing down), but occasionally the period abruptly changes to a smaller value (pulsar instantaneously assumes a higher rotation rate) and the sequence can continue from there. When averaged over several minutes or so, these complexities disappear, leaving only extreme regularity.

These capabilities are important identifying unique characteristics obtained for each pulsar. Thus these are easily seen as a keystone location to determine a spacecraft's locations within a four dimensional space.

Chapter IV

A Tutorial to Solve the 'Free' Two-Body Celestial Mechanics Problem

Abstract: The 'captured' 2-body Kepler problem usually requires bodies with a large difference between their separate masses. The larger body is usually assumed centrally located that influences a smaller satellite body. In binary pulsar orbits, the two bodies may have similar masses as 'free' bodies to generate separate orbits. These bodies with similar weights may produce highly elliptical trajectories while other pulsar binaries with different weights produce near circular orbits; clearly this behavior is counter-intuitive. Consequences of these orbits are either premature or the neutron star may alter gravity due to excessive axis rotation. Closed-form solutions are presented for the 'free' two-body problem. Results indicate the eccentricities for binaries may be different with each of these 'free' orbits. Moreover, there is a correlation that relates eccentricities, versus the mass of the binaries, to the type of trajectories as well as a function of the neutron star's rotation rate. If this is the case, there is a possibility angular momentum may play a role in gravitation.

Keywords: Two-body, celestial mechanics, barycenter, pulsar, integral equations, stability, elliptical, and orbit.

I. INTRODUCTION

When an engineering student normally learns about the 'captured' 2-body problem [1]-[3], there are several restrictive assumptions for the analysis. One basic assumption is there is a large difference between the weights of the two bodies. Moreover, the larger body, say the Earth is immovable and the second lighter body moves similar to a small satellite in orbit around the Earth. These solutions result in an orbit defined as either a circular, elliptical, or a hyperbolic orbit depending upon the satellite's mass, kinetic energy and initial conditions. Accuracy for these problems is rather straightforward and results are satisfactory to predict satellite orbits with these assumptions. Changes to the analytical solution for the satellite-Earth problem can be varied to consider more detailed information regarding the gravitational effects on the Earth. For example, the sphere of the Earth can consider mass density variations due to changes in the gravitational attraction, the presence of a mountain range, an iron ore deposit, the ocean or other effects as a function of the surface location on a spherical or pear-shaped model.

The problem of interest is to evaluate effects within a binary pulsar. Based upon an analysis by Murad [4], a tabulation was formulated to provide information about performance on several binary pulsar orbits. These binary pulsars usually have a neutron star and a companion body or another neutron star that results in separate distinctive elliptical or circular orbits. The problem is the two bodies have relatively similar masses traveling in different or disconcerting orbits centered about a common point of the elliptical orbit's foci points. Clearly this makes the 'free' celestial mechanics problem more complicated. Based upon the knowledge of the first problem, the uneducated would make an initial assumption the pole or focal point of the elliptical orbit must have another gravitational effect or possibly contain a third body. The effort in this paper is to explain how to gain some important insights about the orbits and find a relationship between the masses of the body, rotation rate of the neutron star and the resulting trajectories or their eccentricities.

II. DISCUSSION

Based upon the masses of the neutron star and its companion in a binary pulsar, astronomers have made some judgments about such orbits. However, their judgments may be counter intuitive when the masses are the same as the two bodies, one would expect the orbits to be circular and when the masses are greatly different, they would expect it to be highly elliptical, parabolic or even hyperbolic. In reality, they are found most likely highly elliptical orbits for nearly equal masses in lieu of producing circular orbits while bodies with mass differences produce near circular orbits.

These consequences based upon the conventional wisdom are contrary or counterintuitive toward these expectations. A possible rationale is that the weights are wrong. Another is these differences are based upon gravity, in addition to having a Newtonian attraction, which might have an impact upon angular momentum per Jefimenko [5]-[6], as well as the notions from Winterberg [4] who suggests highly rotating rates of a body, for example a neutron star, could reduce gravitation. These significances suggest there is a coupling effect between the rotation rate of the neutron star and the masses between the two bodies to explain these differences in their orbits. This analysis obviously assumes the initial masses are correct. Thus, there has to be a balance within the neutron star's high rotation rate and these trajectories.

III. ANALYSIS

Before looking at specific orbits and orientations, several notions are required. Trajectories dependent upon several bodies seem to obey a particular law in lieu of resulting in significant collisions. For example, two orbits where the bodies are close to each other, say one orbit is clockwise while the other is counter clockwise, it becomes obvious the two bodies should be attracted to result in a collision. Thus there have to be some assumptions for this analysis.

An initial assumption will be both orbits are in the same motion plane so the problem can be reduced to a two-dimensional analysis. One major assumption is based upon the notion if gravity does not have only a gravitational attraction but produces angular momentum by considering that both orbits must be either clockwise or both counter-clockwise. If this does not occur, the momentum at each orbit results in both a radial and azimuthal force based upon the gravitation from the other body. This becomes a difficult geometric situation that would leave the orbits ending in a collision. These assumptions are discussed in this section.

A. Standard Terminology

The basic problem of the 'captured' two-body model is one body is relatively light in terms of mass while the other body has a significant mass in the same plane. With this premise [1]-[3], the body with the larger mass is assumed to be immovable compared to the first body. The issue is to determine the initial momentum conditions and energy conservation problem essentially based upon the premise the lighter body performs the dynamics that are of interest since the larger body is assumed stationary. The coordinate system uses a polar coordinate system where a unique point is determined based upon a vector length, which has a distance and an angular orientation to completely specify the coordinate location related to a reference coordinate origin. The distance between the two bodies is from a center of the reference coordinate system to include a focal point on the smaller mass' orbit. The center of this reference coordinate system is assumed to also be some negligible distance from the center of the larger body. In reality, there is some small distance treated as inconsequential between the actual locations for the barycenter.

The radial and angular momentum equation for the smaller body is defined as:

$$\overline{F}_r = m\overline{a}_r = m\left(\frac{d^2r}{dt^2} - r\left(\frac{d\theta}{dt}\right)^2\right)\hat{r},$$

$$\overline{F}_\theta = m\overline{a}_\theta = m\left(r\frac{d^2\theta}{dt^2} + 2\frac{dr}{dt}\frac{d\theta}{dt}\right)\hat{\theta} = \frac{m}{r}\frac{d}{dt}\left(r^2\frac{d\theta}{dt}\right)\hat{\theta}.$$

(1)

The subscripts in the LHS are not derivatives but actually the radial and azimuthal force directions respectively. Derivatives are functions of time. The radial force is based upon the gravitational attraction between the two bodies. Moreover, the second equation assumes the azimuthal force vanishes for each of these bodies.

In this problem, the radial dimension is changed as the difference between the distances to the two objects. The problem can be reduced to one dimension with some definitions where μ is the total mass of both bodies ($G\ (m_1+m_2)$). This is considered as the gravitational attraction for this problem as follows:

$$\left(\frac{d^2 r}{dt^2} - r \left(\frac{d\theta}{dt} \right)^2 \right) = - \frac{\mu}{r^2}, \tag{2}$$

$$m \left(r \frac{d^2 \theta}{dt^2} + 2 \frac{dr}{dt} \frac{d\theta}{dt} \right) = \frac{m}{r} \frac{d}{dt} \left(r^2 \frac{d\theta}{dt} \right) = 0., \quad h = r^2 \left(\frac{d\theta}{dt} \right).$$

Clearly the azimuthal gravitation disappears with a constant, h, that is the angular momentum per unit mass used to satisfy this equation for the azimuthal acceleration. Thus the second equation vanishes. At this point, a variable is selected based upon an inverse function of the radius to simplify the problem and removing the time derivatives with substitutions from the problem. This results in:

$$r = \frac{1}{u}, \quad \frac{d\theta}{dt} = hu^2, \quad \frac{dr}{dt} = -\frac{1}{u^2} \frac{du}{d\theta} \frac{d\theta}{dt} =$$

$$= -h \frac{du}{d\theta}; \quad \frac{d^2 r}{dt^2} = -h \frac{d^2 u}{d\theta^2} \frac{d\theta}{dt} = -h^2 u^2 \frac{d^2 u}{d\theta^2}. \tag{3}$$

When these are substituted into the above equation for the radial momentum, the results are:

$$-h^2 u^2 \frac{d^2 u}{d\theta^2} - h^2 u^3 = -\mu u^2, \tag{4}$$

Or with some simplifications:

$$\frac{d^2 u}{d\theta^2} + u = \frac{\mu}{h^2} \tag{5}$$

The solution of this ordinary differential equation considering a geometric length l and eccentricity e is:

$$u = \frac{\mu}{h^2} + C \cos(\theta - \theta_o) \quad \text{or} \quad r = \frac{l}{(1 + e \cos\theta)}. \tag{6}$$

The importance of this equation is the eccentricity e plays a significant role. Basically the smaller body rotates about the larger body with a circular orbit (if e is zero) or if the eccentricity is positive and less than 1.0, the orbit is elliptical with the major body located at one of the focal points in the elliptical orbit. If eccentricity is greater than 1.0, the orbit is hyperbolic and it leaves or escapes the gravitational pull of the larger body. Obviously, this result depends upon initial velocity conditions and kinetic energy before the interaction.

B. 'Free' Body Orbits

The concern is the mass fractions between the 'free' bodies are so different, what occurs if the masses or weights are comparable and how does this impact the trajectories of both bodies? Figure 1 shows a typical situation where if it were possible, the two bodies would have to include an azimuthal gravitational force. The only allowable situation would be if the bodies were moving away from each other at high enough speed but even here, the final orbit would also become questionable. The other possibility is a larger third body may exist at the central focal point. The worry is to define some conditions about stable orbits.

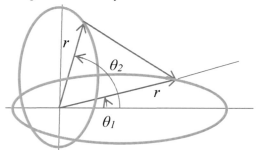

Figure 1. If the two bodies are oriented, they will undergo azimuthal acceleration toward each other. Here the desire of the bodies may leave the orbits and directly move toward each body possibly resulting in a collision.

C. Unstable Orbits

There are several things to destroy this methodology or require considerable complications where the problem is no longer solvable. For example, let us assume the two bodies move in separate elliptical orbits. Under these circumstances, the bodies would be as in Figure 1 and potentially change these elliptical orbits to result in a collision. When crossing orbits, the probability increases for a collision as well as orbital changes due to gravitational overload that traverses on one side of the elliptical orbit over the other. The fact the rotation rate is the same using rotation of the line between the two bodies also implies there is some possibility angular momentum is a component of gravitation. This is specified by Jefimenko and an effort by Lavrentiev et al [7] suggests all of the moons in the solar system operate in the same rotational direction that face the same portion of each moon toward their main planets. Thus, we only see the same side of the Moon while on the Earth. This is true throughout the solar system. Moreover, it is highly probable rotation of each mass can be established in a binary pulsar system. The counter-argument is each of these moons has a gravitational offset to allow this phenomenon.

Let us look at what might appear to be a reasonable stable situation. Place both orbits at the same azimuth orientation as shown in Figure 1. An unstable situation may occur if the bodies are in an orientation that depends upon the separation distance as $r_1 - r_2$, which results in the simplest form of the radial direction in the form:

$$\left(\frac{d^2 r_1}{dt^2} - r_1 \left(\frac{d\theta_1}{dt} \right)^2 \right) = -\frac{\mu_1}{(r_1 - r_2)^2},$$

$$\left(\frac{d^2 r_2}{dt^2} - r_2 \left(\frac{d\theta_2}{dt} \right)^2 \right) = -\frac{\mu_2}{(r_1 - r_2)^2} \quad (7)$$

where $\mu_1 = G m_2$ and $\mu_2 = G m_1$.

And in the simplest form of the azimuthal direction:

(8)

$$m_1 \left(r_1 \frac{d^2 \theta_1}{dt^2} + 2 \frac{dr_1}{dt} \frac{d\theta_1}{dt} \right) = \frac{m_1}{r_1} \frac{d}{dt} \left(r_1^2 \frac{d\theta_1}{dt} \right), \quad h_1 = r_1^2 \left(\frac{d\theta_1}{dt} \right),$$

$$m_2 \left(r_2 \frac{d^2 \theta_2}{dt^2} + 2 \frac{dr_2}{dt} \frac{d\theta_2}{dt} \right) = \frac{m_2}{r_2} \frac{d}{dt} \left(r_2^2 \frac{d\theta_2}{dt} \right), \quad h_2 = r_2^2 \left(\frac{d\theta_2}{dt} \right).$$

In these equations, the *h* values are no longer constants but rather complex geometric relations. However, let us just examine the radial terms. These radial equations reveal when the satellites can cross each other or asymptotically reaches the same radius length. The gravitational attraction when the bodies approach each other will become a singularity or infinite gravitation attraction may result in a possible collision. This obviously is an unstable orbit.

D. *Stable Orbits*

Let us assume both bodies will move in the same rotation direction simultaneously either as clockwise or counter-clockwise. This assumes Jefimenko is probably correct that gravity has a radial attraction similar as Newton as well as produces angular momentum. We will also assume both bodies move with mixed orbits as described in Figure 2 where one is circular and the other is elliptic. Moreover, the focus is at the closest point to the orbit at its closest approach (e.g., perigee).

D.1. The Barycenter Requirement

The issue is if a line between the two bodies at the barycenter coincides with a focal point or not. Obviously this determines the types of orbits. In the initial problem of the captured 2-body problem, the barycenter is located close to the focal point of the orbit. For this issue, can a mixed orbit exist where you have one body in an elliptical orbit and the second body at a circular orbit as seen in Figure 2? The mathematics for the latter are greatly simplified but if they exist, the aerial area per Kepler's law might be violated. This raises another set of conditions if you have elliptical orbits where their long axis are, say 45 or 90 degrees apart as shown in Figure 1. In this case, the bodies at specific locations would require excessive speed to meet these requirements. Looking at Figure 3, the intuitive wisdom is the heavier body should be located at the central location or have a smaller radius as it asymptotically approaches the captured 2-body problem.

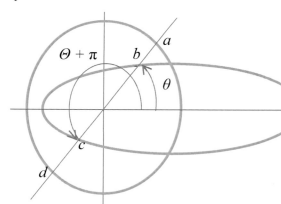

Figure 2. This involves an elliptical and circular orbit combination. Both orbits use common focal point. If the bodies are located in *a* and *b*, they are unstable because they can easily leave the orbits and lead to a collision. If, however, they are located at say, *a* and *c*, or *b* and *d,* the two orbits may represent a stable set of orbits.

Let us assume there are two masses alone in the Universe. They form a system with a center of mass. That center of mass is without acceleration and moves uniformly along some velocity vector – an unchanging velocity vector. We can choose our coordinate inertial system to have its origin anywhere as long as it moves uniformly without any acceleration with constant velocity. We can, therefore, choose our inertial system's origin at one of the two bodies and have an inertial-frame origin move at exactly the constant velocity of the body at *a particular instant of time*. This velocity vector and the observer (sitting on the chosen mass) at the *instant of time* define a plane (from geometry a point and a line defines a plane – the line being the instantaneous velocity vector in this particular case). The usual idea is to select the center of mass, CM, of the two masses (barycenter) to act as the center of coordinates or origin of the coordinate system. Thus a three dimensional framework to a single two-dimensional plane is collapsed. We are in a two-dimensional plane inertial framework moving uniformly alone in the Universe. In essence, the binary orbits remain in one inertial plane moving without rotation and with constant velocity through the universe. There are no other external forces to move any mass out of the plane.

Figure 3. These are representative stable orbits for different orientations. Stability occurs if the line of the two bodies is joined by a focal point that is common to each orbit. In the second case, an unstable orbit occurs at the stars closest to each other if each axis is not collinear.

If the observer is on a mass (of any value relative to the other mass), then this mass exerts a gravitational force on the other mass. The observer on this mass calculates the motion relative to the other body (mass) based upon the inverse-square-law gravitational force and the initial condition (instantaneous velocity vector) of the other mass --- this mass finds it to be a conic section as the solution to the classical two-body problem. This is the result based upon the forces acting against the barycenter. Let us suppose they are at a distance r_1 for m_1 and r_2 for m_2. Again by definition of the CM, the distances must always maintain the same ratio, that is: $r_1 / r_2 = m_2 / m_1$ and this defines the CM defined so a large mass must be closer to the CM fulcrum than a small mass located further to maintain a balance. And as time marches on we can watch the masses approach and receive from us in unison. Thus we have a simple 1D linear coordinate system if no other forces are involved as shown in equations 5 and 6.

Thus the barycenter is an important constraint. Orbits would have to be orientated to fall in the same line for the long axis of the elliptical orbits as well. If there are mixed orbits or elliptical orbits oriented at an angle, then there is a strong possibility the two bodies are influenced by either a third or fourth body(s), a region with unusual gravitational fields representing a possible singularity, or there are solar wind-like effects where particles shower the two bodies with a directional bias. If not, then these orbits are not stable. Thus we shall assume the barycenter is unmovable without the presence of other forces.

D.2. Elliptical Orbits

Let us place the bodies in a two-dimensional plane inertial framework moving uniformly alone in the cosmos. In essence, the pulsar binary orbits remain in one inertial plane moving without rotation and with constant velocity throughout this universe. There are no other external forces to move any mass out of this plane. This reduces as in the previous, a three-dimensional situation to a two-dimensional plane.

Let us assume further that the two bodies move in the same rotational direction with two different elliptical orbits in the same plane. Additionally, let us assume there is a line between both bodies, which also crosses through the focal point and barycenter common to either elliptical or circular orbits. If the line is such, then the distance between both bodies are equal to $r_1 + r_2$ where the first radius is from the first body to the foci and the second radius is the second body to the same foci which is collinear on the same line as the first body. Thus, the distance between the two bodies is easy to establish. Let us assume the angular orientation is at an azimuth angle at θ for the first body where the orientation for the second body is located at $\theta + \pi$ or *180* degrees further in a counter-clockwise direction as a perihelion. Let μ_1 be $G\, m_2$ and μ_2 be $G\, m_1$ to account for the mass terms for gravity.

The governing radial and azimuthal equations for these two bodies now become:

$$\left(\frac{d^2 r_1}{dt^2} - r_1 \left(\frac{d\theta_1}{dt} \right)^2 \right) = -\frac{\mu_1}{(r_1 + r_2)^2},$$

$$\left(\frac{d^2 r_2}{dt^2} - r_2 \left(\frac{d\theta_2}{dt} \right)^2 \right) = -\frac{\mu_2}{(r_1 + r_2)^2}, \qquad (9)$$

$$\left(r_1 \frac{d^2 \theta_1}{dt^2} + 2 \frac{dr_1}{dt} \frac{d\theta_1}{dt} \right) = \frac{1}{r_1}\frac{d}{dt}\left(r_1^2 \frac{d\theta_1}{dt} \right) = 0., \quad h_1 = r_1^2 \frac{d\theta_1}{dt},$$

$$\left(r_2 \frac{d^2 \theta_2}{dt^2} + 2 \frac{dr_2}{dt} \frac{d\theta_2}{dt} \right) = \frac{1}{r_2}\frac{d}{dt}\left(r_2^2 \frac{d\theta_2}{dt} \right) = 0., \quad h_2 = r_2^2 \frac{d\theta_2}{dt}.$$

These are nonlinear equations regarding radial momentum. On this basis, if there is a relationship between the two bodies, whether they both move in elliptical or circular orbits, the common focal point must fall along the line between the two points.

Using the above methodology in equations 1-6, the equations to be solved are as follows:

$$\frac{d^2 u_1}{d\theta_1^2} + u_1 = \frac{\mu_1 u_2^2}{h_1^2 (u_1 + u_2)^2} \quad \text{and} \quad \frac{d^2 u_2}{d\theta_2^2} + u_2 = \frac{\mu_2 u_1^2}{h_2^2 (u_1 + u_2)^2}, \qquad (10)$$

$$\text{where} \quad u_1 = 1/r_1 \quad \text{and} \quad u_2 = 1/r_2.$$

This also influences some relationships with h_1 and h_2. Clearly these are coupled nonlinear equations where one orbit clearly depends upon the second orbit and vice versa. The solution has to have periodic initial and boundary conditions. Note the angle for the second orbit simultaneously includes 180 degrees. This is as follows:

$$u_1 = C_1 \cos(\theta - \theta_o) + \mu_1 \left\{ \int_{\theta_o}^{\theta} \frac{u_2^2(\xi)}{h_1^2(u_1(\xi) + u_2(\xi))^2} \sin(\theta - \xi) d\xi + \right.$$

$$+ \int_{\theta_o}^{\theta_o + 2\pi} \frac{u_2^2(\xi)}{h_1^2(u_1(\xi) + u_2(\xi))^2} \frac{\cos\theta \cos(\xi - \pi/2)}{2\sin\pi/2} d\xi +$$

$$\left. + \int_{\theta_o}^{\theta_o + 2\pi} \frac{u_2^2(\xi)}{h_1^2(u_1(\xi) + u_2(\xi))^2} \frac{\sin\theta \sin(\xi - \pi/2)}{2\sin\pi/2} d\xi \right\}, \text{ and}$$

$$u_2 = C_2 \cos(\theta + \pi - \theta_o') + \mu_2 \left\{ \int_{\theta_o'}^{\theta} \frac{u_1^2(\xi)}{h_2^2(u_1(\xi) + u_2(\xi))^2} \sin(\theta + \pi - \xi) d\xi + \right.$$

$$+ \int_{\theta_o'}^{\theta_o' + 2\pi} \frac{u_1^2(\xi)}{h_2^2(u_1(\xi) + u_2(\xi))^2} \frac{\cos(\theta + \pi) \cos(\xi - \pi/2)}{2\sin\pi/2} d\xi +$$

$$\left. + \int_{\theta_o'}^{\theta_o' + 2\pi} \frac{u_1^2(\xi)}{h_2^2(u_1(\xi) + u_2(\xi))^2} \frac{\sin(\theta + \pi) \sin(\xi - \pi/2)}{2\sin\pi/2} d\xi \right\} \quad (11)$$

where $\theta_o' = \theta_o + \pi$.

Obviously, these are complex nonlinear Volterra Integral equations. Assumptions can be made based upon the relationships between the two different distances and rotation rates to define the *h* terms based upon the mass fractions. Moreover, the rotation rates should also be the same. The latter is a limiting assumption. These are:

$$\frac{r_1}{r_2} = \frac{\mu_1}{\mu_2} = \frac{u_2}{u_1}, \text{ and } \frac{h_1}{h_2} = \frac{r_1^2 \frac{d\theta_1}{dt}}{r_2^2 \frac{d\theta_2}{dt}} = \frac{\mu_1^2}{\mu_2^2}. \quad (12)$$

The second and third integrals result in a constant term. The resulting problem is elliptical orbits about each other:

$$u_1 = C_1 \cos(\theta - \theta_o) + \mu_1 \left\{ \mu_1^2 \int_{\theta_o}^{\theta} \frac{1}{h_1^2} \sin(\theta - \xi) d\xi + \right.$$

$$+ \frac{\mu_1^2}{2\sin\pi/2} \int_{\theta_o}^{\theta_o + 2\pi} \frac{1}{h_1^2} \cos\theta \cos\left(\xi - \pi/2\right) d\xi + \quad (13)$$

$$\left. + \frac{\mu_1^2}{2\sin\pi/2} \int_{\theta_o}^{\theta_o + 2\pi} \frac{1}{h_1^2} \sin\theta \sin\left(\xi - \pi/2\right) d\xi \right\}, \text{ and}$$

$$u_2 = C_2 \cos(\theta + \pi - \theta_o') + \mu_2 \left\{ \mu_2^2 \int_{\theta_o'}^{\theta} \frac{1}{h_2^2} \sin(\theta + \pi - \xi) d\xi + \right.$$

$$+ \frac{\mu_2^2}{2\sin\pi/2} \int_{\theta_o'}^{\theta_o' + 2\pi} \frac{1}{h_2^2} \cos(\theta + \pi) \cos\left(\xi - \pi/2\right) d\xi +$$

$$\left. + \frac{\mu_2^2}{2\sin\pi/2} \int_{\theta_o'}^{\theta_o' + 2\pi} \frac{1}{h_2^2} \sin(\theta + \pi) \sin\left(\xi - \pi/2\right) d\xi \right\}.$$

Adding some more changes results in:

$$u_1 = C_1 \cos(\theta - \theta_o) + \frac{\mu_1^3}{h_1^2} \left\{ \int_{\theta_o}^{\theta} \sin(\theta - \xi) d\xi + \right.$$

$$\left. + \frac{\cos\theta}{2\sin\pi/2} \int_{\theta_o}^{\theta_o + 2\pi} \cos\left(\xi - \pi/2\right) d\xi + \frac{\sin\theta}{2\sin\pi/2} \int_{\theta_o}^{\theta_o + 2\pi} \sin\left(\xi - \pi/2\right) d\xi \right\}, \text{ and} \quad (14)$$

$$u_2 = C_2 \cos(\theta + \pi - \theta_o') + \frac{\mu_2^3}{h_2^2} \left\{ \int_{\theta_o'}^{\theta} \sin(\theta + \pi - \xi) d\xi + \right.$$

$$\left. + \frac{\cos(\theta + \pi)}{2\sin\pi/2} \int_{\theta_o'}^{\theta_o' + 2\pi} \cos\left(\xi - \pi/2\right) d\xi + \frac{\sin(\theta + \pi)}{2\sin\pi/2} \int_{\theta_o'}^{\theta_o' + 2\pi} \sin\left(\xi - \pi/2\right) d\xi \right\}.$$

Note that a constant term appears for all of the integral equations based upon the initial conditions for the two masses. If dealing with a satellite moving about the Earth where μ_1 is very small, the integral expressions for the first orbit has a considerably different value for the multipliers of the second integral.

Integrating these terms using the second angle added by 180 degrees results in:

$$u_1 = C_1 \cos(\theta - \theta_o) + \frac{\mu_1^3}{h_1^2}[1 - \cos(\theta - \theta_o)], \text{ and} \quad (15)$$

$$u_2 = C_2 \cos(\theta + \pi - \theta_o) + \frac{\mu_2^3}{h_2^2}[1 - \cos(\theta + \pi - \theta_o)].$$

The solution assuming the initial angle is at zero degrees results in:

$$r_1 = \frac{1}{+\left(C_1 - \frac{\mu_1^3}{h_1^2}\right)\cos\theta + \frac{\mu_1^3}{2h_1^2}}, \text{ and}$$ (16)

$$r_2 = \frac{1}{+\left(C_2 - \frac{\mu_2^3}{h_2^2}\right)\cos(\theta+\pi) + \frac{\mu_2^3}{2h_2^2}}.$$

Where the constant terms are defined for the initial radius at an initial angle as subscripts measured from the coordinate reference system, r is the initial distance and the subscript is for a particular orbit with:

$$C_1 = +\frac{\mu_1^3}{h_1^2} + \left[\frac{1}{r_1^o} - \frac{\mu_1^3}{2h_1^2}\right]\frac{1}{\cos\theta_o}, \text{ and}$$ (17)

$$C_2 = +\frac{\mu_2^3}{h_2^2} + \left[\frac{1}{r_2^o} - \frac{\mu_2^3}{2h_2^2}\right]\frac{1}{\cos(\theta_o+\pi)}.$$

Combining these terms yields:

$$r_1 = \frac{2h_1^2/\mu_1^3}{1 + \left(\frac{2h_1^2}{\mu_1^3 r_1^o} - 1\right)\frac{\cos\theta}{\cos\theta_o}}, \text{ and}$$ (18)

$$r_2 = \frac{2h_2^2/\mu_2^3}{1 + \left(\frac{2h_2^2}{\mu_2^3 r_2^o} - 1\right)\frac{\cos(\theta+\pi)}{\cos(\theta_o+\pi)}}.$$

Finally, the eccentricities for each of these orbits are:

$$e_1 = \left(\frac{2h_1^2}{\mu_1^3 r_1^o} - 1\right)\bigg/\cos\theta_o, \text{ and}$$ (19)

$$e_2 = \left(\frac{2h_2^2}{\mu_2^3 r_2^o} - 1\right)\bigg/\cos(\theta_o+\pi) = \left(\frac{2h_1^2\mu_2}{\mu_1^4 r_2^o} - 1\right)\bigg/\cos(\theta_o+\pi).$$

Note for situations where mass fraction is almost the same, the result is driven by differences in the initial angles for the different orbits. This results in the two-body problem regarding elliptical, circular or hyperbolic trajectories but with two separately moving bodies. Astronomers call out only a single eccentricity for both bodies in a binary. There is no real reason why one would assume this especially if the masses are not the same; hence, we are showing two different values for each trajectory. If these values are the same, one could have a direct relationship ratioing the h values as a function of the initial radius at each celestial body. Thus the second term shows if these relationships were the same, they would only be true for the same mass fraction terms. Hence there should be different eccentricity for each trajectory. In any extent, the normal convention would have never established these different eccentricities in these trajectories unless a more complete trajectory data evaluation is performed as here.

IV. RESULTS

The question is if any suitable results could be established. Using a portion of the pulsar paper [4] by the author, some data reveals binary pulsars as follows.

Table I. Pulsar Information

Pulsar	Eccentricity e	P(ms)	$M_{primary}$	M_{comp}	Rotation Rate	Orbit Period
8. PSR B1913+16	0.61713	37.904	1.4414	1.3867	26.3Hz	.3229
13. PSR J1903+0327	0.437	2.15	1.74	1.0	465Hz	95.17d

These are shown in Figure 4. The eccentricities of both of these binaries are elliptical and one should expect to see such a trajectory. In a sketch of the first, these orbits are clearly oriented along the long axis. Note the first sketch shows the bodies are directly opposite to each other. One would argue the two bodies are fairly similar in terms of total mass. This tends to substantiate the values shown by looking at the length between the two bodies and measuring the distance to the foci of the elliptical orbits. The lengths are respectively μ_1 and μ_2. This value looks very close to being about .40 or 40%.

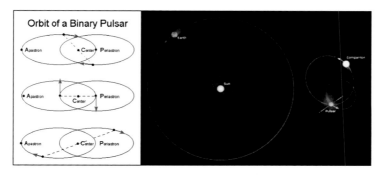

Figure 4. The orbits of the binary pulsar 1913+16 and a comparison of the orbits of the pulsar J1903+0327 with its possible sun-like companion star with the orbit of the Earth around the sun [11].

In Figure 5, assuming this is an adequate representation of the above shown in Figure 4, the μ value varies from this graphic interpretation with no numerical information of .367 to .38. The value here is .364 or about 0.83% different or less than 1 percent. Obviously this result could be altered to reach a similar value. Also note the similar point of view where the longitudinal axis of both orbits is collinear. Moreover, the larger body is located on the smaller or inside trajectory. This last binary is closer to obtaining circular motion where this also implies that when a highly eccentricity orbit exists, it could represent an immature scheme where the orbit. Here the orbit did not 'settle' down and was still in the making additional corrections. Moreover, neither of these orbits has a trajectory which has a major axis at an angle similar to Figure 1 but like the second drawing shown in Figure 3.

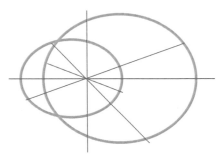

Figure 5. This is a breakdown of the above with J1903+0327. Look at the relations between the distances from the focal point to the separate bodies and the total length between the bodies to form a mass relationship of the bodies.

Early in the discussion, the point was raised about angular momentum effects. In a previous reference [4], the author used Winterberg's ideas vis-a-vis the rotational effect of the neutron star which would decrease its gravitational attraction to the companion. The approach is to alter the gravitational attraction to include some normalized quantity that includes a rotational rate term (ω^2) for the gravitational attraction to both bodies. In this format, the trajectory definition, mass fraction and neutron star rotation rate may provide some correlations directly to establish the value of eccentricity.

These inputs will need:

$$\mu_1 = Gm_2 \quad \text{and} \quad \mu_2 = Gm_1\left(1 + \frac{\omega^2}{2\pi G \rho_c}\right). \tag{20}$$

These terms for the mass fractions would be adjusted to account for the normalization of the angular rotation rate of the neutron star. Of these values, G is the gravitational constant, ρ_c is the gravitational density of the companion and ω is the rotational rate of the neutron star. Note the neutron star rotation rate will drive eccentricity toward an asymptotical circular orbit. This may be the rationale that demonstrates a correlation with gravity and angular momentum.

V. CONCLUSIONS

The basic requirements to identify conditions for creating a binary pulsar indicate the two bodies will have a common focal point at the barycenter, which is expected under the conventional wisdom. Moreover, the orbits will have the largest axis of an elliptical orbit collinear for both bodies. These bodies will rotate both in the same direction either as clockwise or counter-clockwise that by itself implies some angular momentum component as part of gravity. A simple graphic solution indicates the mass ratio between the two bodies could be established. These conclusions support the counter intuitive situation about the choice of masses for binary pulsars may indeed be valid. Thus these general orbits appear to be premature in most of these situations to require the continuous need for normalization of their orbits with extensive time.

By using the approach defined above, separate orbits are defined with separate eccentricities for each of the binary pulsar bodies. These results also indicate there is a coupling between the neutron star's rotation rate and the type of trajectory results based upon mass fraction. This was the original intention of this analysis. If we look at this further, we can only see *causes* but not the direct *effects*. If these are really mature orbits, then these masses may be incorrect based upon angular momentum as a consequence of gravity, which is altered due to the spinning neutron star impact. Thus we may never really measure these masses and their *effects* since we cannot measure these *causes*.

There is something else of importance. There is a harmony that exists for a binary pulsar considering the weights of the neutron star and its companion, the orbit of the two bodies in terms of eccentricity and finally, the neutron star's rotation rate. The celestial mechanics have been reduced in a very simple relationship.

Acknowledgment

The author appreciates comments by his mentor concerning astrodynamics, Dr. R. M. L. Baker, jr. and an incisive review from Morgan Boardman.

References

[1] Baker, Jr. R. M. L.: Astrodynamics- Applications and Advanced Topics, Academic Press, New York and London, 1967.
[2] Battin, R. H.: An Introduction to the Mathematics and Methods of Astrodynamics, ISBN 0-930403-25-8, AIAA Education Series, 1987.
[3] Greenwood, D.T.: Principles of Dynamics, Prentice-Hall, Inc., 1965.
[4] Murad, P. A., Understanding Pulsars to Create a Future Space Propulsor, AIAA Paper at Aerospace Sciences Meeting, Nashville, Tennessee, 2012.
[5] Jefimenko, O. D., Gravitation and Cogravitation, Electret Scientific Company Star City, ISBN 0-917406-15-X, 2006.

[6] Jefimenko, O. D., <u>Electromagnetic Retardation and Theory of Relativity</u>, Electret Scientific Company, Star City, West Virginia, 1997.
[7] Lavrentiev, M. M., Dyatlov, V. L., Fadeev, S. I., Kostova, N. E. and Murad, P. A.: "Rotation Effects of Bodies In Celestial Mechanics", *Proceedings of the Science 2000 Congress*, St. Petersburg, Russia, July 3-8, 2000. Co-authors are from Novosibirsk, Russia.

Nomenclature

a = Reference length
e = Eccentricity
F = Thrust or force
g = Gravity
G = Gravitational constant
m = Mass
r = Radial coordinate
t = Time

Greek Symbols
ρ = Density
ω = Rotation rate
θ = True anomaly
μ = Mass fraction
Subscripts
o = Initial value
$1, 2$ = Individual bodies

Chapter V.

Relativistic Orbit for the 2-Body Celestial Mechanics Problem

Abstract: The 'captured' 2-body Kepler problem considering relativity is solved by using an iterative integral equation. This can be altered to treat the 'Free' 2-body problem associated with a binary pulsar. The rationale for this approach, however, is to increase the accuracy of the limits for a satellite's motion and potentially provide a test to evaluate different gravitational laws. Moreover, this effort might provide additional insights to resolve other anomalies such as the flyby anomaly, the Faint Young Sun paradox, the Pioneer anomaly and other inconsistencies that potentially may be used to validate Einstein's Theory of Relativity. The mathematical solution reveals a correction factor for treating a given closed orbit. However. This correction factor is not a constant value but rather a function of the elliptical or circular orbit angular displacement. This function may be insignificant during portions of the trajectory, say at apogees or perigees. Nonetheless, these results are encouraging where relativity effects may or may not exist for an avenue to understand and resolve several of the other variances currently related to our solar system. For the binary Pulsar issue, rotation as well as an asymmetric gravity tensor warrant a more elaborate evaluation than what is presented here.

PACS: 04.50.Kd, 04.80.Cc, 06.30.Dr, 06.30. Gv, 97.10.-q, 97.10.Gz, 97.10.Xq, 97.60.Gb, 97.60.Jd, 97.60.Lf, 97.80.-d.
Keywords: Binary Pulsar, Neutron Star, Asteroid, Gravity, Jefimenko, Rotation, Angular Momentum, Trajectories.

I. INTRODUCTION

Although the major focus is upon pulsars, the question is whether an effort is worth examining what happens with extremely small values during spacecraft trajectories or not? Once achieved, these results may be extrapolated directly into the Pulsar problem to be considered as an anomaly. Iorio [1] provides an excellent examination of some specific anomalies that are worth examining. To paraphrase and include some of these views, there are currently accepted laws of gravitation applied to known bodies which may have the potential of paving the way for remarkable advances in fundamental physics. One specific gravity law is Winterburg's rule as well as Jefimenko's gravity and co-gravity. This is particularly important now more than ever, given that most of the Universe seems to be made of unknown substances dubbed Dark Matter and Dark Energy or, by contrast, there is a different need for defining a gravitational law other than Newtonian gravitation. Moreover, investigations in one of such directions can serendipitously enrich and find other solutions as well.

The current status [1] of some of these alleged gravitational anomalies in the Solar system are:

a) Possible anomalous advances of *planetary perihelia*,

b) Unexplained orbital residuals of a recently *discovered moon of Uranus* (Mab),

c) The lingering unexplained secular increase of the *eccentricity of the orbit of the Moon*,

d) The so-called *Faint Young Sun Paradox*,

e) The secular decrease of the mass parameter of the Sun,

f) The *Flyby Anomaly*,

g) The *Pioneer Anomaly*, and

h) The anomalous secular increase of the astronomical unit.

In addition to these as well as understanding the orbit dynamics for a binary Pulsar, one more anomaly should also be added to include the Trojan asteroids discussed by the author in [2] which may offer an opportunity for demonstrating the existence of gravitational waves as well as gravitational repulsion.

With these thoughts regarding anomalies, the equations of motion for the two-body celestial mechanics problem, although well-established, is altered. Additionally, several references [3]-[4] have provided supplementary factors to include changes in the trajectories with the Theory of Relativity. One wonders about the magnitude and the impact to these trajectories if for a probe of interest, moving at conditions for an elliptical orbit, which would exert some changes due to relativity. Moreover, the worry should also focus upon trajectory changes if the probe is moving at or near the speed of light. However, we cannot deal with this latter problem at this time. Under these circumstances, light speed trajectories may alter gravitational forces themselves as well as treating only with relativity.

II. DISCUSSION

Let us explain some of these anomalies to gain a better insight observing some unusual events in the cosmos. Basically anomalies [1] may show up in experiments to make comparisons with the conventional wisdom. In science, the word 'anomaly' designates some sort of discrepancies with respect to an expected path observed in a given phenomenon. In astronomical contexts, it was used since ancient times to indicate irregularities in motions of celestial objects. First of all, it must be carefully ascertained if the anomaly really exists; it may be either a mere artifact of the data reduction procedure, or the consequence of malfunctioning of the measuring devices and systematic errors in the observations. Once determined that the anomaly is real, honest engineers and scientists want to determine how to change the conventional wisdom for explaining the anomaly as well as still accept and preserve the existing body of knowledge. Let us address some of these anomalies [1] with the objective that there may be a serious need to change the kinematic motions to explain some of these events.

Anomalous Secular Increase of the *Eccentricity of the Moon's Orbit* demonstrated steady progresses in reviewing data from the Lunar Laser Ranging (LLR) technique. In the last decades, data has determined the orbital changes at a cm level of accuracy or better which allows for accurate testing of the General Theory of Relativity. Moreover, a major limiting factor in our knowledge of the celestial course of the Moon is currently based by a description of the complex geophysical processes. Lingering unexplained increase of the eccentricity of the Moon's orbit are yet to be understood, despite recent efforts to improve the geophysical models of the intricate tidal phenomena taking place in the interior of our planet and its natural satellite.

The eccentricity rate $\dot{e}$ can vary from a high of $1.6 \pm 0.5 \times 10^{-11}$ yr^{-1} to a low of $9 \pm 3 \times 10^{-12}$ yr^{-1}. The general relativistic Lense-Thirring acceleration induced by the Earth's gravitomagnetic field acting on the Moon has the correct order of magnitude, but it does not affect e. A still undetected distant planet in the Solar system does, in principle, make e cumulatively change over time, but the required mass and distance for it are yet to be determined.

For this particular problem, there is a necessity to establish more accuracy about the trajectories for a binary Pulsar. This type of instrumentation technology is currently not existing.

Anomalous perihelion precession of Mercury of 42.98 "cy^{-1} (a change of its orbit by 42.98 arc seconds in a century) since it is nowadays fully included in the state-of-the-art models of all of the modern ephemerides. Instead, if real, it would be due to some unmodeled dynamical effects which, in principle, could potentially signal a breakthrough with the currently accepted laws of gravitation. The relativistic dynamical models for the modern ephemerides, for example Mercury, are not complete that do not include the 1PN gravitomagnetic field of the Sun, not to say of the other major bodies of the Solar system, which causes the Lense-Thirring effect. This effect would be comparable with the action of a hypothetical ring of undetected moonlets in its neighborhood as a possible solution using conventional gravitational physics regarding the gravitational anomalies for Uranus.

For Pulsars, these values are insignificant compared to the accuracy established for the Moon. The precessions for the different rotation rates of the neutron star and a companion may have some unusual connections that warrant improved understanding.

The *Faint Young Sun Paradox*: According to established evolutionary models of the Sun's history, the energy output of our star during the Archean, from 3.8 to 2.5 Gyr ago, would have been insufficient to maintain liquid water on the Earth's surface. Instead, there are strong but compelling independent evidences where our planet was mainly covered by liquid water oceans, hosting also forms of life, during that remote era. As such, our planet could not be entirely frozen during such an era, where it would have necessarily been if it received only about 75% of the current solar irradiance. One view implies a steady recession of the Earth's orbit during the entire Archean eon provided a closer location to its present day heliocentric distance in such a way when the Sun's luminosity was adequate. Thus the effects of the ocean may be a potential gravitational anomaly.

Direct energy from the neutron star and the subsequent reflection of the companion star may provide more information. The dynamics may be of interest holding how the neutron star evolved and how it eventually captured the companion body.

As *'Flyby Anomaly'* is intended to treat the collection of unexplained increases for v_∞ in the asymptotic line-of-sight velocity in the direction of v_∞. This has been of the order of $\approx 1 - 10$ mm s^{-1} with uncertainties to as little as $\approx 0.05 - 0.1$ mm s^{-1}, which have been experienced by the interplanetary spacecraft Galileo, NEAR, Cassini, Rosetta and, perhaps, Juno during their Earth flybys. The flyby anomalies has not yet been detected when such spacecraft flew by other planets. This perhaps may be due to their still relatively inaccurate gravity field models compared to the Earth's gravitational model.

For the solar system, this would give us insights regarding gravitational laws and accuracies. Unfortunately this is not possible for these bodies other than requiring more information as a considerable long function of time.

Pioneer Anomaly: At the end of the twentieth century, it was reported that radio tracking data from the Pioneer 10 and 11 spacecraft exhibited a small anomalous blue-shifted frequency drift uniformly changed the rate of $5.99 \pm 0.01 \times 10^{-9}$ Hz s^{-1} interpreted as a constant and uniform deceleration approximately directed towards the Sun. This was at heliocentric distances approximately of $20 - 70$ au. Each satellite moved in opposite direction from the sun to determine if other effects such as the solar wind would exist.

Subsequent years witnessed some options to explain a variety of conventional and exotic physical mechanisms for both gravitational and non-gravitational nature for these differences. These gravitational effects started when the instrumentation was stopped and the electrical power from a nuclear isotope power supply was altered in a different electrical circuit involving a heater to dissipate electrical energy. In 2012, an appropriate model of the recoil force assumed that an anisotropic emission of thermal radiation off the spacecraft was able to accommodate for about 80% of the unexplained acceleration plaguing the telemetry of both the Pioneer probes as far as magnitude, temporal behavior, and direction of concern. The remaining 20% still does not represent a statistically significant anomaly in view of uncertainties in the acceleration estimates using Doppler telemetry and thermal models. On the other hand, the Pioneer anomaly may be due to some exotic gravitational mechanism external to the spacecraft. This resulted in the form of a constant value and uniform acceleration directed towards the Sun. These views were performed with systematic investigations about its presumed effects on bodies other than the Pioneer probes performed since 2006.

It turned out the Pioneer anomaly may also involve induced anomalous signatures of Uranus, Neptune and Pluto. This would be far too large to consider the initial conditions or strong tensions between the Galactic tide dominant in making Oort cloud comets observable. The action may be a putative Pioneer anomaly-like acceleration in those remote peripheries of the Solar system.

Thus, these anomalies [1] regarding the standard behavior of natural and artificial systems within the Sun's realm as expected may consider where the conventional physics possesses a great potential to uncover modifications of our currently accepted picture of natural laws. Nonetheless, before this dream really comes true, it is mandatory that the unexpected patterns are confirmed to an adequate level of statistical significance by independent analyses, and any possible conventional viable mechanism could be responsible can be reliably excluded.

There are some interesting points concerning neutron stars which may impact a binary Pulsar. According to [5], there was a recent ground-breaking discovery regarding the appearance of powerful jets from a neutron star. This result is based upon the NASA's Chandra Space Telescope where X-ray shock waves originated at the end of jets. The binary system of two stars called Circinus X-1, orbiting each other are about 20,000 light years away from Earth, or about halfway across our Milky Way Galaxy.

While consuming material from a normal companion star torn away from the neutron star's gravitational pull, they claim the neutron star blasts super-heated matter into space along narrow channels called jets that eventually collide with denser gas. The heated collision by blast waves from the jets the colliding with the interstellar gases radiate brightly, from radio wavelengths all the way to X-rays.

This reference [5] infers: "Some theories suggest that they are made by tapping into the rotation energy of a black hole, similar to a giant flywheel that stores energy. In the case of the black hole, this energy is stored in a giant vortex of space-time that is constantly dragged around the black hole. Neutron stars have powerful jets similar to black holes, but there is no vortex effect, so something else must be powering the jet."

By contrast, there is a problem in that the rotation of the neutron star could generate this material which is created by the debris from either creating the supernova or from consuming the body from the companion star. With its strong gravitational pull, the rotation can cause the jet to be similar as frame-dragging. The spiral could also result from the movement of the spinning neutron star. Likewise the mass in the jet can also create as another gravitational source that may impact the neutron star and the companion star.

In terms of black holes, the jets are claimed to be created by mass in the accretion disk. The matter in the disk is in tension between the black hole's gravitational pull and the centrifugal motion of the debris in the disk. Since either a black hole or a neutron star can be created by a supernova, similar debris could exist. What would be interesting is to find a jet without a rotating black hole. Such a jet would not have an accretion disk and the material would be due to solely the black hole. The speed of the jet should demonstrate 'natural' particles moving faster than the speed of light. However, as mentioned by Jefimenko and Winterburg, the probability of finding a none-rotating black hole is nonexistent because of the dynamics in the cosmos.

In previous studies by the author [6], a Green's function solution was treated for the trajectories in a binary pulsar. The interesting factor within the discipline of astrophysics, states these systems are usually identified with a single eccentricity value for both celestial bodies. These results [6] indicate for most of these binary systems, the trajectories may not be duplicates of each other for a neutron star and its companion. There is some situations, which could result where one body is in a near circular orbit while the other companion or neutron star is clearly in an elliptical orbit. Such situations for these orbital trajectories demand separate eccentricities.

Previous results of this assessment suggests the different eccentricities for binary pulsars between a neutron star and a companion should be based upon:

$$e_1 = \left(\frac{2h_1^2}{\mu_1^3 r_1^o} - 1\right)\bigg/\cos\theta_o, \text{ and}$$

$$e_2 = \left(\frac{2h_2^2}{\mu_2^3 r_2^o} - 1\right)\bigg/\cos(\theta_o + \pi) = \left(\frac{2h_1^2 \mu_2}{\mu_1^4 r_2^o} - 1\right)\bigg/\cos(\theta_o + \pi).$$

(1)

These differences are a function of several parameters such as angular momentum, initial orbital parameters as well as weight or gravitation.

III. ANALYSIS

Before looking at specific orbits and orientations, several notions are required. This will be discussed for a basic understanding shown with typical orbit definition, a Green's function, the impact of relativity to the trajectory, and finally a solution to the integral equation for relativity trajectory.

a. *Standard Terminology*

A solution for the 2-body problem previously mentioned is an integral equation using a Green's function solution accounting for boundary conditions is:

$$u(\theta) = C_1 \cos(\theta - \theta_o) + \frac{\mu}{h^2}\left\{\int_{\theta_o}^{\theta} \sin(\theta - \xi) d\xi + \frac{\cos\theta}{2\sin\pi/2}\int_{\theta_o}^{\theta_o + 2\pi} \cos\left(\xi - \pi/2\right) d\xi \right. \quad (2)$$

$$\left. + \frac{\sin\theta}{2\sin\pi/2}\int_{\theta_o}^{\theta_o + 2\pi} \sin\left(\xi - \pi/2\right) d\xi\right\}.$$

These terms are the solution for the above equation where the Green's function will become the kernel of the integral equation which is defined as:

$$u(\theta) = C_1 \cos(\theta - \theta_o) + \alpha K(\theta, \theta), \quad (3)$$

$$\text{where: } K(\theta, \theta) = \left\{\int_{\theta_o}^{\theta} \sin(\theta - \xi) d\xi + \frac{\cos\theta}{2\sin\pi/2}\int_{\theta_o}^{\theta_o + 2\pi} \cos\left(\xi - \pi/2\right) d\xi \right.$$

$$\left. + \frac{\sin\theta}{2\sin\pi/2}\int_{\theta_o}^{\theta_o + 2\pi} \sin\left(\xi - \pi/2\right) d\xi\right\}$$

This is the basic solution to the problem without relativistic effects.

b. *Relativistic Mechanics*

Relativistic effects can vary the sense of time dilation and changes in length. Such changes depend upon the velocity. Let our probe move at a stationary orbit about the Earth. The probe's trajectory can be given for a geodesic [7] in:

$$\frac{d^2 x^\alpha}{d\tau^2} + \Gamma^\alpha_{\beta\gamma} \frac{dx^\beta}{d\tau} \frac{dx^\gamma}{d\tau} = 0. \quad (4)$$

where τ is the proper time. This is rewritten as:

$$\frac{d}{d\tau}\left(g_{\lambda v}\frac{dx^v}{d\tau}\right) - \frac{1}{2}\frac{\partial g_{\mu v}}{\partial x^\lambda}\frac{dx^\mu}{d\tau}\frac{dx^v}{d\tau} = 0. \quad (5)$$

The proper time is based upon the space-time interval that depends upon the metric:

$$ds^2 = -c^2 d\tau^2 = g_{\mu v} dx^\mu dx^v. \quad (6)$$

Considering the Sun's gravitational potential having a mass of M, the Schwartzschild metric using standard coordinates in a spherical coordinate system is:

$$ds^2 = -c^2 d\tau^2 = \left(1-\frac{2GM}{c^2 r}\right)^{-1} dr^2 + r^2 \left(d\theta^2 + \sin^2\theta\, d\phi^2\right) - \left(1-\frac{2GM}{c^2 r}\right) c^2 dt^2. \quad (7)$$

After considerable terms and assumptions related to a plane based upon Newtonian Mechanics and defining constants of integration, this becomes:

$$\left(\frac{h}{r^2}\frac{dr}{d\theta}\right)^2 = c^2(k-1) + \frac{2GM}{r} - \frac{h^2}{r^2}\left(1-\frac{2GM}{c^2 r}\right). \quad (8)$$

If you allow $u = 1/r$ as previously and obtain $dr/d\theta = -r^2 du/d\theta$ as in the original definition, this results into:

$$\left(\frac{du}{d\theta}\right)^2 + u^2 = c^2\frac{(k-1)}{h^2} + \frac{2GM}{h^2}u + \frac{2GM}{c^2}u^3. \quad (9)$$

Differentiating this equation results in:

$$\frac{d^2 u}{d\theta^2} + u = \frac{GM}{h^2} + 3\frac{GM}{c^2}u^2 = \alpha - \beta u^2. \quad (10)$$

Note the value of β is basically from the Theory of Relativity and is small for the Earth-satellite situation which is why this effect is not usually considered regarding short-term celestial mechanics. However, do we fully understand the impact of this value with respect to a given trajectory? Moreover, for a neutron star, the impact of rotation may also be required and alter this value compared to the Earth. Let us consider this point.

c. Solution Rationale

Using relativity, the integral equation for the above ordinary differential equation is of interest as:

$$u(\theta) = C_1 \cos(\theta - \theta_o) + \alpha K(\theta,\theta) - \beta \int_0^\theta K(\theta,\xi) u^2(\xi) d\xi =$$
$$= \varsigma(\theta,\theta) - \beta \int_0^\theta K(\theta,\xi) u^2(\xi) d\xi .$$
(11)

This is an inhomogeneous Friedholm equation or a Volterra integral equation. Because of the squared term[4] for the independent variable and the coupling between these terms, this is nonlinear. The first term is a previously determined orbit trajectory solution without relativity. Here, we are assuming this tends to minimize the coupling impact with the integral equation.

Normally in using an iterative process, an initial equation (the first term) is assumed as a starting position. This is then used in the next correction using the recent previous values and the process continues until you can identify specific terms that represent solutions. The process is used differently here where the initial assumed value is the orbital trajectory solution without relativity. In this fashion, this is similar to evaluating a perturbation to the results using the nonlinear effects for corrections. The first term on the RHS represents using an iterative process to start the solution. This process is performed in a similar iterative fashion; however, some terms will not be included because of orders of magnitude effects. This means for generating a series, if it exists, the absolute magnitude of the kernel is basically less than the value of 1 and with numerous K^n terms, this becomes insignificant.

This would be used to asymptotically approach the answer. The process for u starts with u_0 with the first iterative and so on as u_1 on up to u_2 and u_3:

$$u_0(\theta) = \varsigma - \beta \int_0^\theta K(\theta,\xi) u_0^2(\xi) d\xi , \quad or$$
$$u_1(\theta) = \varsigma - \beta \int_0^\theta K(\theta,\xi) \left[\varsigma - \beta \int_0^\theta K(\theta,\xi) u_0^2(\xi) d\xi . \right]^2 d\xi .$$
(12)

$$then: u_1(\theta) = \varsigma - \beta K(\theta,\theta) \varsigma^2 + 2\varsigma \beta^2 K \int_0^\theta K(\theta,\xi) u_0^2(\xi) d\xi + ...$$
(13)

Using this as another iteration, this becomes:

---

[4] It is interesting to note where the form of this equation may also solve a problem in fluid dynamics as well as with other disciplines. Solving this problem, therefore is of value. Furthermore, the approach of using integral equations offers additional tools resolving mathematical physic challenges.

$$u_2(\theta) = \zeta - \beta K(\theta,\theta)\zeta^2 + \qquad (14)$$
$$+ \zeta \beta^2 K \int_0^\theta K(\theta,\xi)\left[\zeta - \beta K(\theta,\theta)\zeta^2 + 2\zeta \beta^2 K \int_0^\theta K(\theta,\xi)u_1^2(\xi)d\xi + ...\right]^2 d\xi + ...$$

Let: $\varphi = \beta K \zeta$, then the next iteration results in:

$$u_3(\theta) = \frac{1}{\beta K}\left[\varphi - \varphi^2 + 2\varphi^3 - 4\varphi^4 + 2\varphi^5 + \right. \qquad (15)$$
$$\left. + 8\frac{\varphi^3}{K^2}(\varphi - \varphi^2)\int_0^\theta K(\theta,\xi)u_2^2(\xi)d\xi + ...\right.$$

This becomes:

$$u_3(\theta) = \frac{\varphi}{\beta K}\left[1 - \varphi\left[1 - 2\varphi + 2^2\varphi^2 - 2^3\varphi^3 + 2^4\varphi^4 12^5\varphi^5 + ...\right.\right. \qquad (16)$$

Or for the nth iteration, this is a series where:

$$u_n(\theta) = \frac{\varphi}{\beta K}\left[1 - \varphi\left[1 - 2\varphi + 2^2\varphi^2 - 2^3\varphi^3 + 2^4\varphi^4 - 2^5\varphi^5 - .... \pm 2^{n-1}\varphi^{n-1}\right.\right. . \qquad (17)$$

Simplifying this results in:

$$u_n(\theta) = \frac{\varphi}{\beta K}\left[1 - \frac{\varphi}{1+2\varphi}\right] = \zeta\left[1 - \frac{\varphi}{1+2\varphi}\right]. \qquad (18)$$

This can be further simplified in a form to separate the elliptical trajectory results with the objective of creating a correction factor:

$$u(\theta) = \zeta\left[\frac{(1+K(\theta,\theta)) + \beta\alpha K(\theta,\theta)}{(1+K(\theta,\theta)) + 2\beta\alpha K(\theta,\theta)}\right] = \zeta\left[\frac{1+K(\theta,\theta)(1+\beta\alpha)}{1+K(\theta,\theta)(1+2\beta\alpha)}\right] \qquad (19)$$

Note the first term in the RHS is the elliptical trajectory equation without relativity effects. The right-hand term on the RHS represents a correction factor to allow for relativity. The term is general enough to include a Green's function using an initial value or a boundary value problem.

This results in:

$$u(\theta) = \varsigma \left[\frac{1 + K(\theta,\theta)\left[1 - 3\left(\frac{GM}{hc}\right)^2\right]}{1 + K(\theta,\theta)\left[1 - 6\left(\frac{GM}{hc}\right)^2\right]} \right] = \varsigma \psi \quad (20)$$

Thus this includes the original trajectory with a correction factor. The feature ψ is the desired correction factor that, because of the kernel, will not be a constant function but rather a function of angular displacement during the orbit. For this, the integral equation kernel can be used for initial value or boundary condition problems. The kernel in the above problem where the initial angle θ_0 is zero will become:

$$K(\theta,\theta) = \left\{ \int_{\theta_o}^{\theta} \sin(\theta - \xi)\, d\xi + \frac{\cos\theta}{2\sin\pi/2} \int_{\theta_o}^{\theta_o + 2\pi} \cos\left(\xi - \pi/2\right) d\xi + \frac{\sin\theta}{2\sin\pi/2} \int_{\theta_o}^{\theta_o + 2\pi} \sin\left(\xi - \pi/2\right) d\xi \right\} \quad (21)$$

or for these conditions:

$$K(\theta,\theta) = 1 - \cos\theta. \quad (22)$$

And thus the correction factor is:

$$\psi(\theta) = \left[\frac{1 + (1 - \cos\theta)\left[1 - 3\left(\frac{GM}{hc}\right)^2\right]}{1 + (1 - \cos\theta)\left[1 - 6\left(\frac{GM}{hc}\right)^2\right]} \right]. \quad (23)$$

The problem for the Earth is when the numbers are included for a circular stationary orbit, the multipliers involve 11 decimal places of 9 such as .9999999… onwards before different numerical values appear. This is the right-hand terms on the numerator and denominator. Clearly, since we do not have the sensitivity for these values, ψ can easily be assumed to be unity and for all practical purposes, the solution for the two body problem is more than adequate in terms of accuracy to ignore relativity. On this basis, it is clear the problem of Mercury's perihelion required a measure of small minute amounts of azimuthal change over a considerable century to assess the contributions from the theory of relativity. Moreover, additional research may provide some rationales to explain the other anomalies previously mentioned.

For the binary Pulsar issue, the asymmetric gravitational tensor may be required that allows angular motion effects with gravitation. This would occur in equations (3)-(6) changing the trajectory equations.

IV. CONCLUSIONS

The purpose of this effort was to determine a potential means for using a testing function to assess different gravitational laws hopefully with examining a closed-loop trajectory. After considerable mathematics to treat the complexity of this problem, a factor was identified that would provide the trajectory without relativity compared to the same trajectory with relativity. The resulting factor appeared to be a function of the azimuthal direction which occurs during an elliptical orbit; however, this azimuthal effect may be of a small consequence unless there is adequate sensitivity for these effects. This result opens the door for explaining several of the anomalous behavior within the solar system.

References

[1] Iorio, L., "Gravitational Anomalies in the Solar System?", arXiv:1412.7673v1, 21 Dec 2014.
[2] Murad, P. A., "To Find Different Gravity Laws... Proof Wave Equations Are Real and Gravity Generates Gravitational Waves", to be presented at STAIF II, Albuquerque, New Mexico, 2015.
[3] Jefimenko, O. D., Gravitation and Cogravitation, Electret Scientific Company Star City, ISBN 0-917406-15-X, 2006.
[4] Jefimenko, O. D., Electromagnetic Retardation and Theory of Relativity, Electret Scientific Company, Star City, West Virginia, 1997.
[5] Sell, P. H., Heinz, S., Calvelo, D. E., Tudose, V., Soleri, P., Fender, R. P., Jonker, P. G., Schulz, N. S., Brandt, W. N., Nowak, M. A., Wijnands, R., van der Klis, M., Casella, P., "Parsec-Scale Bipolar X-Ray Shocks produced by Powerful Jets from the Neutron Star CIRCINUX X-1", Astrophysical Journal Letters, 2010, p. L194, Draft version December 11, 2013 1008.0647v1 [astro-ph.HE] 3 Aug 2010.
[6] Murad, P. A.,"A Tutorial to Solve the 'Free' Two-Body Binary Pulsar Celestial Mechanics Problem" to be presented at JSE, and presented at STAIF II, Albuquerque, New Mexico, 2013.
[7] R. A. Nelson, Handbook on Relativistic Time Transfer, Edition 3.1, Bethesda, Maryland, Oct 2001.

Nomenclature

a = Reference length
e = Eccentricity
F = Thrust or force
g = Gravity
G = Gravitational constant
m = Mass
r = Radial coordinate
t = Time

Greek Symbols
ρ = Density
ω = Rotation rate
θ = True anomaly
μ = Mass fraction

Subscripts
o = Initial value
$1, 2$ = Individual bodies

Chapter VI.

Propulsion and Implications

Abstract: There are many unknowns concerning stellar evolution of neutron stars. Neutron stars might possess multipolar architecture in lieu of a single dipole claimed by the conventional wisdom. The multipole issue cannot be resolved using a single point observer reference point such as the Earth, but would require a non-terrestrial observer location with a significant offset. Moreover, the question is how a neutron star's magnetic field would be created considering differences between the neutron core and a gas surface layer consisting of protons and electrons. These differences between the layers constitute electrical charges with moving currents in a magnetic field supported by a fast moving rotating core. If electrons in Cooper pairs exist in a neutron star, then the amount of magnetism may increase by acting analogous with superconductivity. By symmetry, proton pairs should also exist to produce similar charge distributions. With these thoughts, a laboratory replica of a neutron star is proposed to create a large magnet suitable for a space mission.

PACS: 04.50.Kd, 04.80.Cc, 06.30.Dr, 06.30. Gv, 97.10.-q, 97.10.Gz, 97.10.Xq, 97.60.Gb, 97.60.Jd, 97.60.Lf, 97.80.-d.
Keywords: Binary Pulsar, Neutron Star, Asteroid, Gravity, Jefimenko, Rotation, Angular Momentum, Trajectories.

I. INTRODUCTION

There are two major concerns regarding neutron stars that might have a propulsion perspective. The first is to use a strong magnet with similar capabilities derived by a neutron star and the second is to investigate if there are influences validating Winterberg's equation using high rotations to reduce the gravitational field. The first might be usable to supply a magnetic field to generate a primary propulsion system or may actually be suitable to create a thrust influence. The notion to create a magnet would warrant a laboratory replication for a neutron star. The other problem is to use the Poynting field to take advantage of the huge amount of mass and rotation within the neutron star.

II. ANALYSIS AND RESULTS

It is not questionable at all about using the presence of a large magnet in space and its impact to induce motion or supplement a space drive. Such a device may take advantage of the solar system's magnetic field and act to oppose its alignment as a dipole with respect to the magnetic flow. If the view that Gertenshtein [1] and Forward [2] are correct where magnetism and electricity relates to gravitation as well as the Murad-Brandenburg Equation [3] concerning the Poynting conservation could induce a torsion or possibly a gravitational field, may allow a propulsion system to operate. The objective is to place such a large magnet in a far-abroad trajectory moving toward celestial bodies toward the outer reaches of the solar system or toward travel trajectories moving toward another star.

A. A Potential Model for a Neutron Star.

With all of these differences, what would typical model in a laboratory be used to create an environment to represent a neutron star? The question about a perfectly spherical core is crucial. In nuclear explosions, the implosion process requires uniformly displaced charges surrounded about the nuclear fuel. Each charge is equal in strength and these are detonated simultaneously to produce an optimal nuclear reaction. If any of these charges fail, chances are high that a jet will squirt outside of the shell and result in reducing the energy of the nuclear detonation. For a supernova, the shells are not evenly distributed and sometimes jets most likely result. The implication is the explosion is not evenly distributed and the possibility of resulting with a spherical neutron star is extremely low.

If a rotating body spins in an axis-symmetric fashion, a vortex will be created oriented along the spinning axis. This vortex would consist of electron and proton gases to generate a magnetic field. Since we do not see a beacon where the magnetic field pole is oriented with an azimuthal offset from the rotational axis, it is safe to assume the neutron core is not perfectly spherical.

Assuming the neutron core creates edges of a neutron crystal or matrix, each of these edges can create hydrodynamic vortices due to the neutron star's weather based upon the rotational rate and gas layer thickness. This atmosphere is not a few protons or electrons but has to be considerable to create large magnetic fields. Electron pairs or Cooper pairs will generate intense magnetism similar to superconductivity. It is amazing electrons and protons have the same magnitude in terms of charge while they have a difference in mass of 1 to 1840. Apparently the electrons because of their lower mass have a higher level of efficiency so to speak. If electron pairs exist, it is reasonable by symmetry that proton pairs can also exist. The layers of the proton and electron gases generate separate charges while their currents, based upon vortices, will induce large magnetic fields which may either support or oppose each other. These vortex filaments made of electrons or protons are created by Taylor instabilities.

These views are different from the conventional wisdom and indicate each pulsar represents a unique capability based upon the protuberances on the surface topology. Moreover, the neutron star may be ovaloid. Although they can be defined generally, the topology covered by the rotation rate as well as presence of a companion or other star(s) further attests to the uniqueness of these cosmic events.

B. *A Magnet to Create a Space Propulsion Scheme*

The propulsion system or magnet is rather simple. It will consist of a large air-bearing device using a spherical ball manufactured with Neodymium and nickel married in a steel alloy. This will have small fins that will act similar to a turbine forcing the ball to rotate within a structural casing. The clearance between the ball and casing will be large enough with spacing clearance to withstand a very large voltage once the ball lifts off of the casing. The voltages would be set at 511 KV using microwaves or by using other means, which are the voltage needed to strip electrons from atoms as well as anomalous behavior as observed by Maker [3], [4].

The gas-pressurized ball bearing will operate a gas with hydrogen. This will dissociate the Hydrogen into electron and proton gases as separate entities. There will be sets of hydrogen jets to induce rotation on the ball concentric with respect to the casing to spin at some speed of say, 600 revolutions per second or higher. The system gas spins the ball and levitates it centered by the casing. When conditions permit, the electrostatic voltage is provided and the magnet should operate as expected. Both the core and casing should act as an electric charge. The moving core with the gas mixture of electrons and protons should create a magnetic field. Moreover, a Faraday cage would be required for biological safety due to the extensive electrical currents and some of the magnetic effects. Obviously the device should be located at some extension to the basic body structure away from instrumentation or any biological functions. A hybrid fusion reactor may be required to generate the electricity.

Other suitable geometries could be used with these magnets. These can conform to specific requirements basic with the space ship and to use specific volume requirements. It is unfortunate that a more realistic geometry such as an ovaloid could be used but the constraint is to generate the ball to lift-off the casing surface.

C. *The Poynting Field and a Gravitational Field*

The author and a colleague [2] looked at the influence of rotation effects and how it could generate a force. This includes the Russian effort by Godin and

Roschin as well as our own Morningstar Energy Box[5]. Basically the latter was an electrodynamic device weighing approximately 190 pounds and when operating under steady-state operation reduced weight by 7%. Under transient situations, the weight would be reduced to as much as 20%.

These electromagnetic devices generally rotated in a manner to create a three-dimensional magnetic and or electrical field. This 3-D motion is generally different from what one would expect with a symmetric electric motor or generator. The major notion was to examine the Poynting field.

When examining the Poynting field, the Poynting vector is the vector cross product between the electric and magnetic field. The first question is if the magnetic and electric fields are wave equations, can the Poynting field also have partial differential equations to generate waves. This was found true. The next issue was to define a relationship to account for rotation. In other words can the curl of the Poynting vector exist as part of a conservation law? During the derivation, this term was amplified to include the curl of the curl term, which indicates there indeed was a correlation concerning rotational effects.

In the process of this derivation, there appeared to be a Cauchy-Riemann-like relationship normally used for elliptical partial differential equations. In this derivation, a similar relationship was pursued to extend the resulting conservation equation to create another field. The question was if this field was an unknown torsion field or a new type of description for a localized gravitational field.

Subsequent results for a Poynting Field conservation law is:

$$\mu_0 \left[\frac{1}{c^2} \frac{\partial^2 \overline{S}}{\partial t^2} - \nabla^2 \overline{S} \right] = -4\pi \left[\rho_m \nabla \times \overline{E} - \rho_e \nabla \times \overline{B} + \frac{1}{c^2} \frac{\partial}{\partial t} \left(\overline{J}_e \times \overline{B} + \overline{E} \times \overline{J}_m \right) \right] + \\ + \mu_0 \nabla \times \nabla \times \overline{S}. \quad (1)$$

This second field as a wave equation was:

$$\frac{1}{c^2} \frac{\partial^2 \overline{V}}{\partial t^2} - \nabla^2 \overline{V} = \frac{\partial}{\partial t} \int_0^r \left[4\pi \left[\rho_m \nabla \times \overline{E} - \rho_e \nabla \times \overline{B} \right] - \mu_0 \nabla \times \nabla \times \overline{S} \right] \cdot dr + \\ + 4\pi \nabla \cdot \left[\overline{J}_e \times \overline{B} + \overline{E} \times \overline{J}_m \right]. \quad (2)$$

This equation as mentioned, may be a more realistic use of generating a gravitational field than others regarding using electric and magnetic fields. Moreover, this may have more granularity than Winterberg with:

$$\nabla \cdot \overline{g} = -4\pi G \rho = 2\omega^2, \quad \text{where} \quad \rho = -\frac{\omega^2}{2\pi G}. \quad (3)$$

---

[5] This approach examined some concerns regarding unusual aerodynamic/electromagnetic phenomenon.

The field expression for a pulsar would treat a steady-state situation considering the high angular rotation rate. This becomes:

$$-\nabla^2 \bar{V} = +4\pi \nabla \cdot \left[\bar{J}_e \times \bar{B} + \bar{E} \times J_m \right]. \tag{4}$$

This unfortunately does not include the curl term but suggests currents and the fields would drive these terms.

If we look at fields, we could talk about creating a Poynting motivator for a candidate propulsor where the force is:

$$\bar{F} = e(\bar{E} + \bar{v} \times \bar{B}) + \frac{1}{4\pi} \bar{E} \times \bar{B} = e\bar{E} + \bar{J}_e \times \bar{B} + \frac{1}{4\pi} \bar{E} \times \bar{B}. \tag{5}$$

This would ignore the first and second terms in the RHS of the equation, which also depends upon the charge and current dependent upon the presence of electrons or ions to produce the charge. Significant amounts of energy for the last term would be required to perform these tasks. Such a motivator, for example, depends strictly upon the Poynting contribution in the last term of the RHS.

III. CONCLUSIONS

Obviously the mechanisms for a neutron star are not usually well understood. These characteristics depend upon masses, rotation, and the presences of jets. Most of these factors have been ignored from the perspective of creating a magnet or for a space drive to go into deep space. The problem is that work and more research is insisted as well as keeping an open-mind, inquisitive mind, and an exploratory interest. This understanding would easily represent an embryonic and surprise technology because of the implications to create green technology.

References

[1] Forward, R. L., "Guidelines to Antigravity," *American Journal of Physics*, vol. 31, pp166-170, 1963.
[2] P. A. Murad and J. E. Brandenburg, "A Poynting Vector/Field Conservation Equation and Gravity- The Murad-Brandenburg Equation," presented in STAIF II, Feb 2012, Albuquerque, New Mexico.
[3] Maker, D., *"Propulsion Implications of a New Source for the Einstein Equations"*, in proceedings of Space Technology and International Forum (STAIF 2001), edited by M. El-Genk, AIP Proceeding Melville, NY, 2001, pp.618-629.

[4] Maker, D., *"Very Large Propulsive Effects Predicted for a 512 kV Rotator"*, IAC-04-S.P.10, 55th International Astronautical Congress of the International Astronautical Federation, the International Academy of Astronautics, and the International Institute of Space Law, Vancouver, Canada, Oct. 4-8, 2004.

Nomenclature

B = Magnetic field intensity (volt-second/m^2)
c = Speed of light (3×10^8 meters/sec.)
E = Electric field intensity (volts/meter)
F = Force (Newtons)
g = Gravity (9.8 m/sec^2)
J = Current (coulomb/sec)
m = Mass (kilograms)
p = Pressure
v = Velocity (meters/second)

Greek Symbols

ε = permitivity (farads/meter)
μ = Permeability (henrys/meter)
σ = Conductivity (mhos/meter)
Λ = Cosmological constant
γ = Relativity factor
ρ = Volume charge density (coulombs/m^3)

Epilogue

The original objective of this monogram was to examine the unusual wonders of the different trajectories within a binary pulsar. Unfortunately, the 'free' body problem was not normally covered during the sixties in celestial dynamics courses compared to the 'captured' two-body problem. This only deals with a small satellite in orbit about a far larger celestial body, which differs in the context where there are two masses having somewhat similar weights. Nonetheless, astronomers assumed that other trajectories would also result in similar fashions. However, strange things exist as anomalies regarding the influence of neutron stars.

The problem is to accept reality where these heavy bodies rotate at significant angular rotation rates, masses which would not easily replicate this phenomenon in a laboratory environment. With everything in life or Mother Nature, there is some sort of unknown harmony within a binary pulsar between the weights of the neutron star, its companion, and the rotation rate. The implication is the gravitational attraction may not be solely an attractive force but some angular momentum effects may exist to create this harmony.

This view implies heresy compared to Newtonian gravity and even the relativistic physicists may also be repelled by this view. The problem is such tools are present to the latter because of the existence of a gravitational tensor. Here, there may occur an off-axis component on this tensor which allows coupling between angular momentum with the gravitational attraction as initially suggested by Jefimenko. We also see this with two relatively young moons on Himalia and Elara in orbit around Jupiter. With one irregularly shaped body and another that is almost spherical are at about the same altitude and seem to rotate at the same rate almost identically.

There is no physical reason for this to occur and even this defies the Newtonian who argues the rotational rate depends upon an offset in the gravitational center of the moon. As suggested, this occurs to all of the main planets with respect to their larger satellite moons. Is this equivalent to argue angular momentum is coupled with gravity? Again, this would clarify the coupling in a binary pulsar between the weights for a neutron star and its companion compared to the neutron star's rotation rate. The harmony of this achievement is provided in a simple expression which solves one of the main riddles of Mother Nature.

One would ask why this would have any relevance. The answer is simple. It first implies our understanding of gravity that may be incomplete and we crave to warrant continual research investigations. Secondly, this effect may have some insights to look at a space propulsion drive to allow us to travel to the far-abroad.

On this basis, my wishes would be this document, with and without its many flaws, might eventually motivate some individual to solve these riddles and to build a conventional wisdom to finally create a meaningful Warp Drive. Such a space capability would allow mankind to satisfy its insatiable curiosity regarding the cosmos. It is also my wishes where the reader may enjoy this effort and finds it useful as well…

P.A. Murad

Subject Index

A
Angular Momentum 44
Anomaly 10, 23, 92
Asteroid 44, 92

B
Barycenter 77, 81
Binary Pulsar 24, 50, 52
Black Hole 17, 30, 45, 95
Brandenburg, J. 54, 68, 106
Buoyancy 65

C
Celestial mechanics 52, 76, 92
Chandra X-ray Observatory 68, 95
Conservation Law 16, 108
Cooper Pair 69, 106
Cosmological Constant 19, 20, 37
Curvature Tensor 19, 27. 32, 38

D
Dyatlov 17, 38, 90

E
Eccentricity 10, 45, 52, 78, 87, 92, 96
Einstein Theory of Relativity 9, 16-20, 36
Elara 44-45
Electric field 30, 108
Electrodynamic 36, 58, 66, 108
Electron 25, 42, 63-66, 106
Electron Gas 63, 65-66, 107
Elliptical 43, 49, 76, 92, 108

F
Flyby Anomaly 10, 92, 94
Forward, R. 68, 71, 106, 109
Friedholm Equation 99

G
Gas Dynamic 25, 27, 46
Gertenshtein 106
Godin 17, 26, 109
Glitching 73
GPS 48, 66
Gravitational Distortion 28, 38
Gravitational Tensor 4, 20, 23, 29, 34, 111
Gravitational vortex 29, 38
Gravity 23, 32, 44, 68, 83, 94

H
Heaviside 42
Himalia 40, 45, 111
Hubble Telescope 4

I
Instability 62
Integral equation 84, 85, 97, 101

J
Jefimenko 23, 42, 44, 50, 80
Jeong 42, 46
Jet 17, 41, 45, 95, 106
Jupiter 41, 44, 66

K
Kalaza-Klein 35
Kepler's Law 81
Kosyrev 17

L
Lavrentiev 79

Lighthouse Effect   49, 51, 64

M
Mach's Principle   18
Magnetic field   24, 47, 58, 65, 107
Magnetar   24, 46, 47
Magnetofluid   58
Maker, D.   70, 107
Mass Ratio   89
Micropulse   73
Maxwell's Equations   22, 32, 65
Multiverse   35
Murad-Brandenburg Equation   68, 106

N
Neodymium   70, 107
Neutron Crystal   69, 106
Neutron Star   48, 52, 57, 62, 68, 88, 106
Nickel   70, 107
Nova   45, 60-62, 96

O
Orbit   10, 82, 99
Ovaloid   60, 63, 107
Over-unity Performance   38

P
Paradox   10, 92, 94
Parallel Universe   16, 29, 38
Physical Vacuum   17, 18
Pioneer 10   59, 60, 94
Pioneer 11   59, 60, 94
Poynting Conservation   106
Propulsion   105
Proton Gas   63, 65, 69, 107
Pulse Drift   73
Pulse Nulling   73
Pulsar   41, 73
Pulsar Timing   48, 63, 68, 73

Q
Quark   67, 68
Quark Star   67, 68
Quadrupole   66
Quasar   65, 67

R
Ricci Curvature Tensor   19, 27-29
Rotation   24, 42-46, 65, 76

S
Sirius Movie   16
Space-continuum   38
Speed of Light   17-19, 29, 42, 54, 96
Stability   82
Stellar Evolution   60, 105
Stress-Energy Tensor   19, 20, 22, 30-32, 34
Superconductivity   69, 106
Super Nova   45
Symmetry Form   32, 33

T
Torsion   17, 68, 106
Torsion Field   17, 18, 22, 38, 108
Trajectories   43, 76, 79, 86, 96

U
Universe   16, 20, 30, 67, 81, 92
Unstable   20, 60, 62, 81

V
Vacuum   18, 20, 28, 35
Vacuum Energy   20
Vis-à-vis   88

Volterra Equation 84, 99
Vortex 29, 38, 66, 96, 106

W
Warp Drive 23, 34
Williams, Pharis 21
Winterburg 92, 96

X
X-ray 42, 58, 67, 95
X-ray Observatory 68
X-ray Pulsar 42, 58

Y
Yilmaz 37
Young Sun Paradox 92, 94

Z
Zero-Point 20
Zeeman Effect 58
ZPF (Zero Point Field) 35, 36

Other books from the author:

"A Magnificent Odyssey" published by 1st Books.com

"The Demon's Gate" published by iUniverse.com
Rewritten book by Amazon.com

Part I of the Demon's Gate Trilogy
"Beyond The Demon's Gate: The Phaeton Affaire" published by Amazon.com

Part II of the Demon's Gate Trilogy
"Beyond The Demon's Gate: The South American Foray" published by Amazon.com

Part III of the Demon's Gate Trilogy
"Beyond The Demon's Gate: Taking The Russian Alien" published by Amazon.com

"The Quest and Wrath of the Gods: A 'Demon's Gate' Sequel" published by Amazon.com

"The Dark Side Of The Tiger" published by Amazon.com

"There is a Crazy Man Living in my Basement" published by Amazon.com

*"The Eagle and the Dragon
- An End of Days Tale from the Book of Revelations"* to be published

Made in the USA
San Bernardino, CA
03 September 2015